H. Fischer, P. Hofmann, F. R. Kreissl,
R. R. Schrock, U. Schubert, K. Weiss

Carbyne Complexes

VCH

E. O. Fischer

H. Fischer, P. Hofmann, F. R. Kreissl,
R. R. Schrock, U. Schubert, K. Weiss

Carbyne
Complexes

VCH

This book was carefully produced. Nevertheless, authors, editors and publisher do not warrant the information contained therein to be free of errors. Readers are advised to keep in mind that statements, data, illustrations, procedural details or other items may inadvertently be inaccurate.

Published jointly by
VCH Verlagsgesellschaft, Weinheim (Federal Republic of Germany)
VCH Publishers, New York, NY (USA)

Editorial Director: Dr. Michael G. Weller
Composition: Hagedornsatz GmbH, D-6806 Viernheim
Printing: Heidelberger Verlagsanstalt und Druckerei GmbH, D-6900 Heidelberg
Bookbinding: Verlagsbuchbinderei Kränkl, D-6148 Heppenheim
Cover design: TWI, H. J. Weisbrod, D-6943 Birkenau

Cover illustration: Structure of $[(C_6H_5C)W(CO)_4I]$, the first metal carbyne complex synthesized by Fischer et al. in 1973.

A Cip catalogue record for this book is available from the British Library

Deutsche Bibliothek Cataloguing-in-Publication Data:

Carbyne complexes : [dedicated to Ernst Otto Fischer on the occasion of his 70th birthday] / Helmut Fischer ... – Weinheim ; New York : VCH, 1988
  ISBN 3-527-26948-7 (Weinheim) Pp.
  ISBN 0-89573-849-X (New York) Pp.
NE: Fischer, Helmut [Mitverf.]; Fischer, Ernst O.: Festschrift

Distribution
VCH Verlagsgesellschaft, P. O. Box 1260/1280, D-6940 Weinheim
    (Federal Republic of Germany)
Switzerland: VCH Verlags-AG, P. O. Box, CH-4020 Basel (Switzerland)
Great Britain and Ireland: VCH Publishers (UK) Ltd., 8 Wellington Court, Wellington Street, Cambridge CB 1 HW (Great Britain)
USA and Canada: VCH Publishers, Suite 909, 220 East 23rd Street, New York, NY 10010-4606
    (USA)

This book is dedicated
to
Professor Dr. E. O. Fischer
on the occasion
of his 70th birthday.

Foreword

Ernst Otto Fischer, scientist and teacher, world citizen and humanist, completed his 70th year on 10th November 1988. He did not, however, equate this with completion, or finality, or the last word. No indeed, always seeking intellectual stimulation through new perceptions, he set a pace so rapid that his peers all over the world sometimes had difficulty keeping up with him. His students and colleagues have now prepared a personal birthday present for him to remind him of years of fulfilment as a teacher and researcher and to acquaint the world of chemistry with an important era in Organometallic Chemistry. The present monograph, dealing with carbyne complexes of the transition metals, provides a most knowledgeable and informative source dealing with the chemistry of the carbon-metal triple bond.

Ernst Otto Fischer, the scientist. If one had to select the gems from the cornucopeia of his scientific achievements the following would stand out amongst the rest: the assignment of the structure of ferrocene in collaboration with W. Pfab (1952)[1]; the synthesis of dibenzenechromium in collaboration with W. Hafner (1955)[2] and the subsequent establishment of the π-aromatic complexes; synthesis of the first metal-carbene complex (with A. Maasböl, 1964)[3]; the synthesis of the first metal-carbyne complex (with G. Kreis, 1973)[4]. The consistent, untiring expansion of these classes of compounds has, however, played an even much greater part than the aforementioned achievements in advancing Organometallic Chemistry and in establishing it as an independent but interdisciplinary field of chemical research. To his undisputed services rendered to the history of chemistry belongs the ability to have developed, from initially surprising single observations, a convincing, tangible and useful system of Organometallic Chemistry for the chemical community. Hundreds, perhaps thousands, of well characterized compounds have found their way into the literature during E. O. Fischer's more than thirty year long 'Research Adventure'. He learned the solid basis of his craft under Walter Hieber (1895–1976), a pioneer in metal-carbonyl chemistry[5], his immediate predecessor in the Chair of Inorganic Chemistry at the then Technische Hochschule München. There, in 1952, E. O. Fischer obtained his doctorate for his work on the synthesis of tetracarbonyl-nickel by reduction of nickel salts with dithionite in aqueous solution[6]. The metal carbonyls were, in the future, under E. O. Fischer, to signalize the breakthrough into the new dimensions in the chemistry of the metal-carbon bond. Together with other workers a rich harvest was gathered in, ranging from a deeper understanding of the nature of the chemical bond in general to the preparative and industrial use of organometallic systems in particular.

VIII

Ernst Otto Fischer, the teacher. "This award is the merit of my coworkers" – this was Fischer's comment when he was awarded the Nobel Prize (jointly with G. Wilkinson) 15 years ago. There before us stood the academic teacher modestly and honourably sharing the fame which he had never pursued! That is how we know and love him, we who have, for a longer or shorter time, been privileged to belong to his circle. We were a family to him, and the spirit of give and take which resulted from this relationship spurred the scientific commitment which led to such far-reaching achievements. Fischer's laboratory made it clear to all what chemistry was all about, namely that it was an intellectual craft. This was clear even to the first year students for whom the director of the institute always had the greatest solicitude. This short recital of the accomplishments of the academic teacher would be incomplete if one were to forget his unique ability to inspire younger scientists. He conveyed to them the joy of making a new substance, of obtaining a correct elemental analysis or – in spite of scepticism – of making a complete spectral characterization ("*One* method is *no* method"; E. O. F.).

He would have liked to study the history of art – the scientist as aesthete. However, just as he later drew his students into the fascination of Organometallic Chemistry, so Walter Hieber introduced the soldier on leave from the front in 1941/42 to chemistry. His successful life has made it difficult for us younger chemists because "es ist nicht immer leicht, nach den Fußstapfen unserer Vorfahren zu wandeln, wenn diese zu lange Beine gehabt haben".[7]

Ad multos annos to the celebrator of this anniversary and to the book!

Wolfgang A. Herrmann
Technische Universität München

1 E. O. Fischer and W. Pfab, Z. Naturforsch. *7B* (1952) 377.
2 E. O. Fischer and W. Hafner, Z. Naturforsch. *10B* (1955) 665.
3 E. O. Fischer and A. Maasböl, Angew. Chem. *76* (1964) 745; Angew. Chem. Internat. Edit. Engl. *3* (1964) 580.
4 E. O. Fischer, G. Kreis, C. G. Kreiter, J. Müller, G. Huttner, and H. Lorenz, Angew. Chem. *85* (1973) 618; Angew. Chem. Internat. Edit. Engl. *12* (1973) 564.
5 E. O. Fischer, Chem. Ber. *112* (1979) XXI. – W. A. Herrmann, Chemie in unserer Zeit **22** (1988) 113.
6 W. Hieber and E. O. Fischer, Z. Anorg. Allgem. Chem. *269* (1952) 292 and 308.
7 Peter Rosegger: "Als ich noch der Waldbauernbub war", pg. 218, Bertelsmann-Verlag, Gütersloh (Germany) 1956. – Translation: "It is not always easy to follow in the footsteps of our ancestors, especially when they have had too long legs."

Preface

This book is dedicated to Professor E. O. Fischer on the occasion of his 70th birthday. The authors have selected for their articles various aspects of the fascinating chemistry of metal carbyne complexes. The book concentrates on mononuclear carbyne complexes with terminally bonded CR ligands. Covering synthesis, electronic and solid state structures, reactivity, and catalysis, it is our aim to present a state-of-the-art report on the chemistry of the metal-carbon triple bond.

H. Fischer
P. Hofmann
F. R. Kreissl
R. R. Schrock
U. Schubert
K. Weiss

List of Contributors

Prof. Dr. Helmut Fischer
Institut für Anorganische Chemie
Universitätsstraße 10
7750 Konstanz
Federal Republic of Germany

Prof. Dr. Peter Hofmann
Anorganisch-chemisches Institut der
Technischen Universität München
Lichtenbergstraße 4
8046 Garching
Federal Republic of Germany

Prof. Dr. Fritz R. Kreissl
Anorganisch-chemisches Institut der
Technischen Universität München
Lichtenbergstraße 4
8046 Garching
Federal Republic of Germany

Prof. Richard R. Schrock
Dept. of Chemistry
Massachusetts Institute of Technology
18-390 Massachusetts Avenue
Cambridge, MA 02130
USA

Prof. Dr. Ulrich Schubert
Institut für Anorganische Chemie
Am Hubland
8700 Würzburg
Federal Republic of Germany

Dr. Karin Weiss
Laboratorium für Anorganische Chemie der Universität
Universitätsstraße 30
8580 Bayreuth
Federal Republic of Germany

Contents

Helmut Fischer
The Synthesis of Fischer-type Carbyne Complexes   1

Ulrich Schubert

Solid State Structures of Carbyne Complexes   39

Peter Hofmann

Electronic Structures of Transition Metal Carbyne Complexes   59

Fritz R. Kreissl
Selected Reactions of Carbyne Complexes 99

John S. Murdzek and Richard R. Schrock

High Oxidation State Alkylidyne Complexes   147

Karin Weiss

Catalytic Reactions of Carbyne Complexes 205

List of abbreviations

Me	methyl	Et	ethyl
Pr	n-propyl	Pr^i	i-propyl
Bu	n-butyl	Bu^i	iso-butyl
Bu^t	tert. butyl	Np	neopentyl
$c\text{-}C_6H_{11}$	cyclohexyl	Cp	η^5-cyclopentadienyl
Cp*	η^5-C_5Me_5	MeCp	η^5-C_5H_4Me
Fc	ferrocenyl	Ind	indenyl
Ph	phenyl	Bz	benzyl
Tol	tolyl	Mes	mesityl
Tos	tosyl	Anis	anisyl
o	ortho	m	meta
p	para	mer	meridional
fac	facial	thf	tetrahydrofuran
py	pyridine	phen	phenanthroline
ophen	phenanthroline	pic	4-picoline
bipy	bipyridyl	dme	dimethoxyethane
dppe	$Ph_2PCH_2CH_2PPh_2$	dmpe	$Me_2PCH_2CH_2PMe_2$
dccd	dicyclohexylcarbodiimide		
tmeda	tetrametylethylenediamine		
ttp	5.10.15.20-tetraphenylporphinato		

The Synthesis of Fischer-type Carbyne Complexes

By Helmut Fischer

1 Introduction

The first synthesis of transition complexes with a formal metal-carbon triple bond was reported by E. O. Fischer et al. in 1973 (1). Since then, the field of transition metal carbyne (alkylidyne) complexes has developed rapidly because of the great interest in the unusual properties of these compounds and their possible significance in organic synthesis and as intermediates in catalytic reactions. Several hundred complexes have been synthesized, isolated and characterized. A number of preparative routes are now available for the synthesis of different types of carbyne complexes. Several reviews have appeared which, however, cover only parts of the area (2–5).

This chapter will concentrate on the synthesis of carbyne complexes of group VI to VIII metals in low oxidation states (Fischer-type complexes). Complexes with "bridging" carbyne ligands are excluded. A separate chapter is devoted to the synthesis of alkylidyne complexes of "early" transition metals and group VI metals in higher oxidation states (Schrock-type complexes). Throughout this chapter which covers the literature up to November/December 1987 the notation carbyne complex is used exclusively.

Different strategies for the preparation of carbyne complexes are possible. New carbyne complexes can be obtained from non-carbyne complex precursors as well as by modification of already existing carbyne complexes. The vast majority of the elaborated procedures may be grouped into one of the following classes:

(A) syntheses from non-carbyne complex precursors

    (a) by transformation of a transition metalbonded sp^2-hybridized carbon atom into a sp-carbyne carbon atom,

    (b) by transformation of a transition metalbonded sp-hybridized carbon atom (e.g. in CS, CNR) into a sp-carbyne carbon atom,

(B) syntheses by modification of already existing carbyne complexes.

2 Syntheses from Non-Carbyne Metal Complexes by Transformation of a Transition Metal Bonded sp² Carbon Atom into a sp-Carbyne Carbon Atom

2.1 Preparation from Carbene Complexes

2.1.1 Reaction with Lewis Acids

The first synthesis of carbyne complexes was accomplished by treatment of alkoxycarbene complexes of chromium and tungsten (Fischer-type carbene complexes) with boron trihalides (1): Scheme 1.

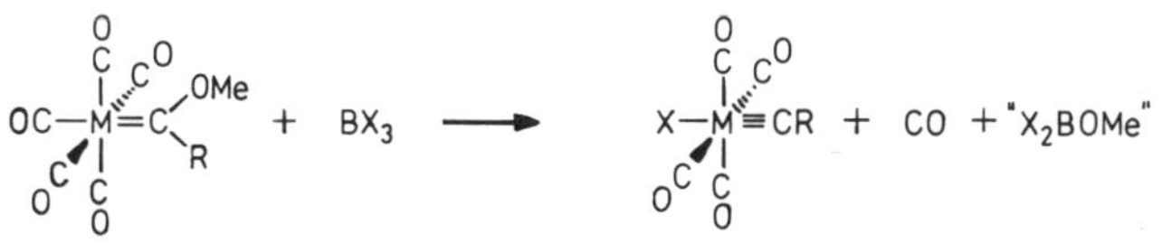

$$OC-\underset{C}{\overset{C}{M}}=C\overset{OMe}{\underset{R}{}} \ + \ BX_3 \ \longrightarrow \ X-\underset{C}{\overset{C}{M}}\equiv CR \ + \ CO \ + \ ^{"}X_2BOMe^{"}$$

M = Cr, Mo, W;   R = Me, Ph;   X = Cl, Br, I.

Scheme 1.

The carbyne complexes were obtained in high yield. This route is not confined to alkoxy(aryl) or -(alkyl)carbene complexes and boron trihalides. A number of different heteroatom stabilized carbene complexes may be used as precursors and different Lewis acids may be employed. The reaction is the most useful and versatile route for the synthesis of carbyne complexes from non-carbyne complexes and probably the majority of neutral carbyne complexes prepared so far were synthesized by this method.

R may be an alkyl (1, 6–10), a cyclopropyl (11) or an aryl group (which may be substituted by donor or by acceptor groups) (1, 6, 12–17), a vinylic (18), an acetylenic (19) or a thienyl group (20), a dialkylamino (21–25), an imino (26, 27) or a silyl group (28, 29).

Even carbene complexes in which R is an organometallic fragment such as $(\eta^5\text{-}C_5H_4)Mn(CO)_3$ (30), $(\eta^5\text{-}C_6H_5)Cr(CO)_3$ (31), $(\eta^5\text{-}C_5H_4)Fe(\eta^5\text{-}C_5H_5)$ (32, 33), $(\eta^5\text{-}C_5H_3Me)Fe(\eta^5\text{-}C_5H_4Me)$ (34), $(\eta^5\text{-}C_5H_4)Ru(\eta^5\text{-}C_5H_5)$ (34) may be used, e.g. Scheme 2.

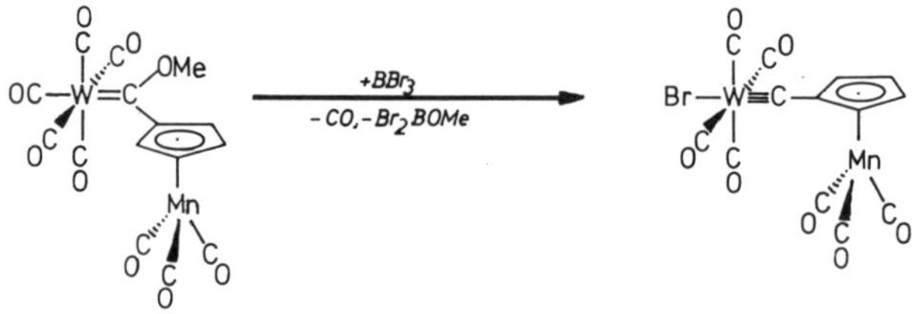

Scheme 2.

Biscarbyne complexes may be obtained from biscarbene complexes (35, 36), e.g. Scheme 3.

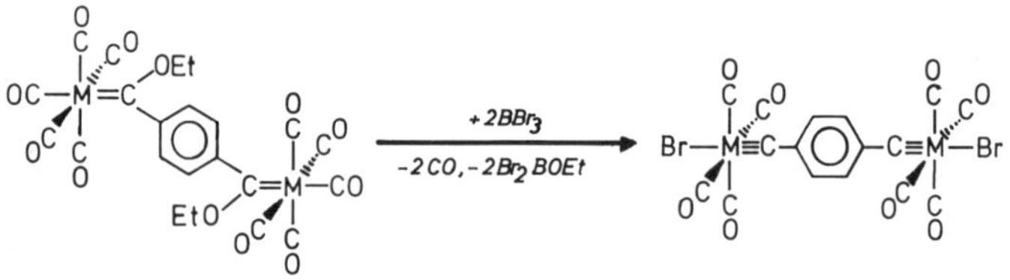

Scheme 3.

The leaving group in the carbene complex precursors (OMe in Scheme 1) may also be varied. Instead of methoxy- or ethoxycarbene complexes, amino- (37, 38), thio- (39), siloxy- (40) or acetoxycarbene complexes (20) can be employed. Most of these precursor complexes are synthesized from alkoxycarbene complexes therefore substituting them for alkoxycarbene complexes does not offer any advantage, moreover, alkoxycarbene complexes generally are (a) easier to prepare, (b) obtainable, in most cases, in very high yield and (c) easy to handle.

Apart from boron trihalides, other Lewis acids such as aluminium or gallium trihalides (28, 41) or ethylaluminium dichloride (41) can be used. However, Et_2AlCl, Et_2AlF and Et_3Al do not react with carbene complexes. In some cases, substituting Al_2Br_6 for BBr_3 considerably improves the yield of carbyne complexes [e.g. (12)]. In general, the thermal stability of the resulting carbyne complexes increases in the series Cl, Br, I.

When substituting BF_3 for BX_3 (X = Cl, Br, I) the reaction takes a somewhat different course. Instead of trans-fluorocarbyne complexes, trans-BF_4-complexes are formed (42–44), e.g. Scheme 4.

The BF_4^- ligand is easily displaced by other nucleophiles such as SCN^-, CN^- (42), $AsPh_3$, $Bu^t\text{-}N \equiv C$ (43) or even H_2O (44).

The first step in the reaction of alkoxycarbene complexes with Lewis acids very likely involves an electrophilic attack of the Lewis acid (LA) at the oxygen

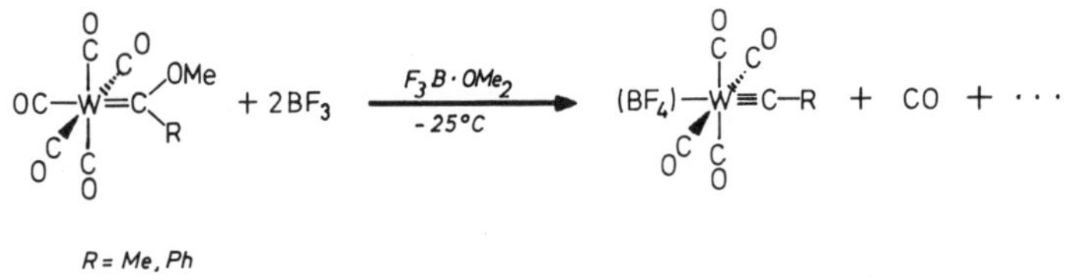

R = Me, Ph

Scheme 4.

atom of the alkoxy group followed by elimination of $(RO-LA)^-$ from the complex to give cationic carbyne complexes. Since the carbyne ligand is a very strong π-acceptor, the stability of the resulting cationic complex depends strongly on the other ligands, in particular on the trans-ligand. If the trans-ligand is a strong π-acceptor like CO the M-(CO) bond is broken and CO is replaced by the halide to give neutral trans-halogeno carbyne complexes. If the trans-ligand is a weak π-acceptor or even a donor group, the cationic carbyne complex may be stable enough to be isolated. This is nicely demonstrated in the reaction of *cis-* and *trans*-$(Me_3Y)(CO)_4Cr[C(OMe)Me]$ (Y = P, As, Sb) with BX_3 (Scheme 5) (44–47).

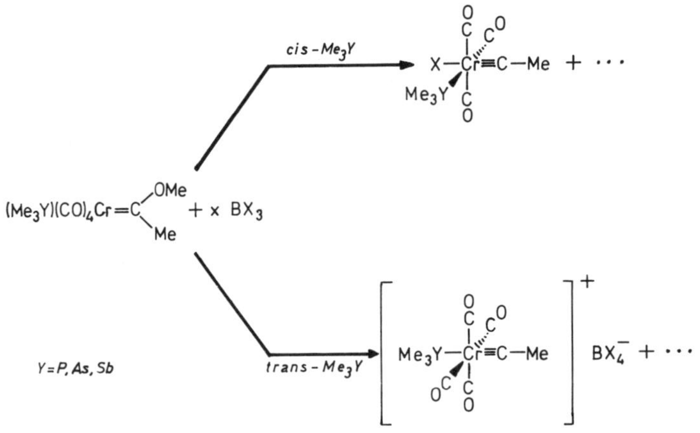

Scheme 5.

In principle, several strategies are possible to stabilize cationic carbyne complexes:
(a) "blocking" of the trans-position by weak π-acceptor or donor groups,
(b) increasing the backbonding properties of the metal-ligand fragment, LL'M, by use of strong π-donor ligands,

(c) increasing the backbonding property of the metal-ligand fragment by use of "late" transition metals (of group VIII or X) in combination with donor substituents,

(d) reducing the π-acceptor property of the carbyne ligand, C-R, by choice of substituents R which are better donors than alkyl or aryl.

Another example of (a), apart from that shown in Scheme 5, is the following reaction (48) (Scheme 6).

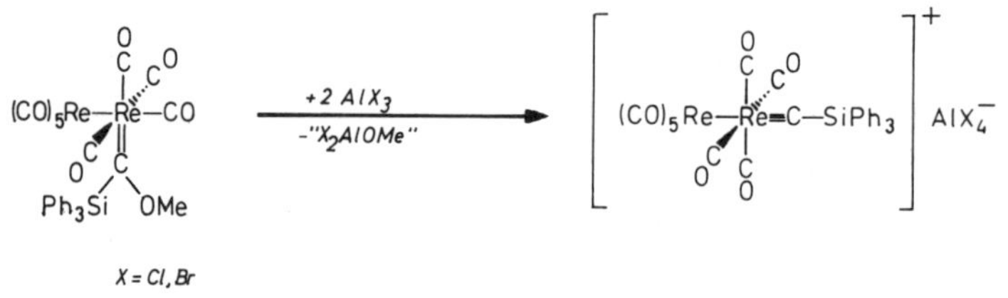

X = Cl, Br

Scheme 6.

The cationic carbyne complex (resulting from AX_3-assisted elimination of OMe^- from the cis-carbene complex) stabilizes through isomerization to give the trans-carbyne complex.

Principle (b) is especially useful for the synthesis of a variety of cationic carbyne complexes of not only manganese (49–55) and rhenium (56, 57) but also chromium (58). The benzene ring in the chromium compounds may also be substituted by methyl groups. Examples are shown in Schemes 7 and 8.

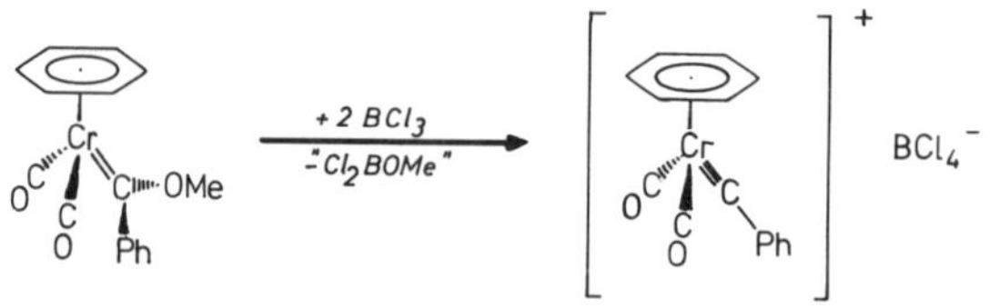

Scheme 7.

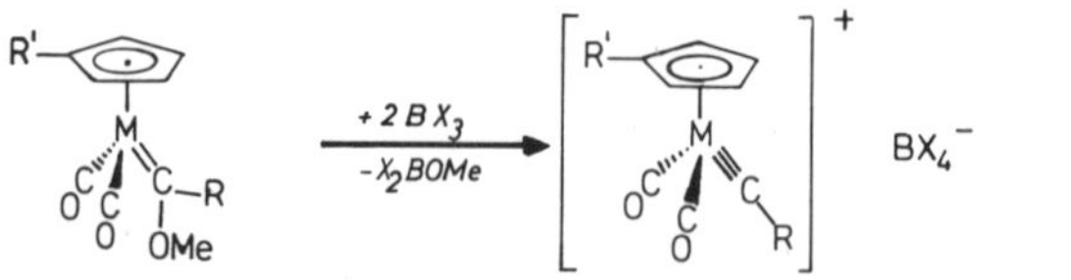

$M = Mn$ $R = Me, Ph, NEt_3, (\eta^5-C_5H_4)Mn(CO)_3$ $R' = H, Me$

$(\eta^5-C_5H_4)Fe(\eta^5-C_5H_5)$

Scheme 8. $M = Re$ $R = Ph, SiPh_3$ $R' = H$

Cationic carbyne complexes of iron (59) and nickel (60) can be obtained by applying principles (e) and (d) (Schemes 9 and 10).

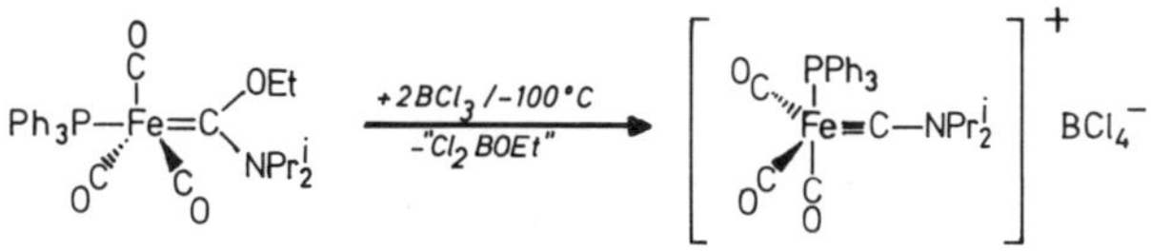

Scheme 9.

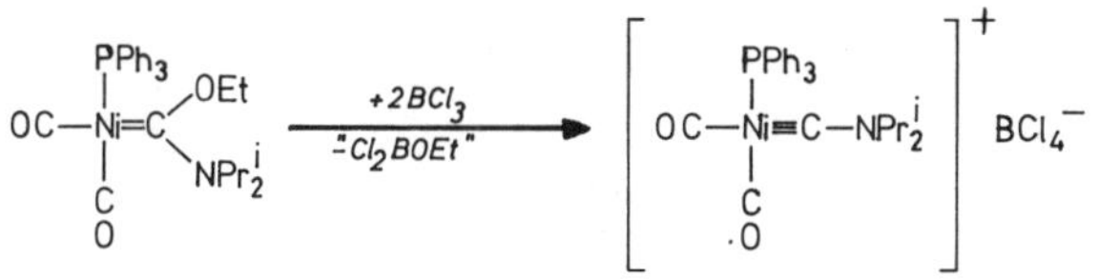

Scheme 10.

The amino substituent on the carbyne carbon atom stabilizes the corresponding complexes considerably through interaction of the lone electron pair of nitrogen with a p-orbital of the carbyne carbon atom (61, 62). Thus it is even possible to synthesize cationic pentacarbonyl carbyne complexes (23, 25, 63–65), e.g. Scheme 11.

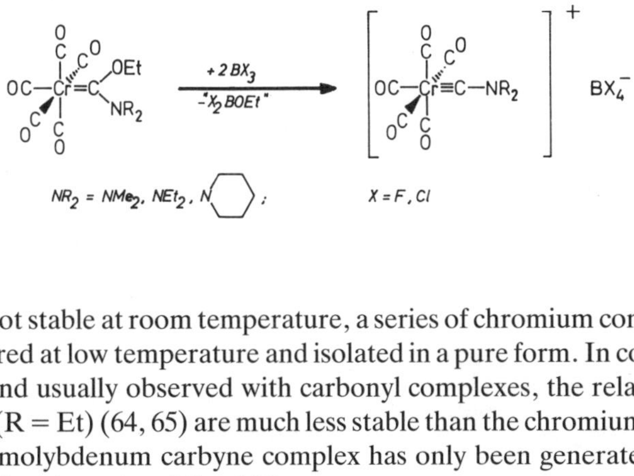

NR_2 = NMe_2, NEt_2, N⟨⟩ ; $X = F, Cl$

Scheme 11.

Although not stable at room temperature, a series of chromium complexes have been prepared at low temperature and isolated in a pure form. In contrast to the stability trend usually observed with carbonyl complexes, the related tungsten complexes (R = Et) (64, 65) are much less stable than the chromium complexes. As yet the molybdenum carbyne complex has only been generated at −78 °C and trapped with nucleophiles (66).

Judged by the results of spectroscopic and X-ray diffraction studies and according to HMO calculations (61, 67) the 2-azoniavinylidene form (B) (Scheme 12) contributes significantly to the overall structure (Scheme 12).

$$\left[(CO)_5M\equiv C-\bar{N}R_2 \right]^+ \quad \longleftrightarrow \quad \left[(CO)_5\bar{M}=C=NR_2 \right]^+$$

(A)   (B)

Scheme 12.

Cationic carbyne complexes are strongly electrophilic and therefore any one of a large variety of nucleophiles can be added rapidly at the carbyne carbon atom. This reaction is very useful for the synthesis of carbene complexes, especially those which are not accessible by any other route [see (68)]. Diarylamino-carbyne complexes cannot be prepared by the method summarized in Scheme 11 since the corresponding diarylaminocarbene complex precursors are not accessible via nucleophilic attack of NPh_2^- on the carbon atom of a coordinated CO in $M(CO)_6$ (due to the low nucleophilicity of NPh_2^-) and subsequent alkylation. This problem can be circumvented by Lewis acid assisted fragmentation of amino(imino)carbene complexes (25, 69) (Scheme 13):

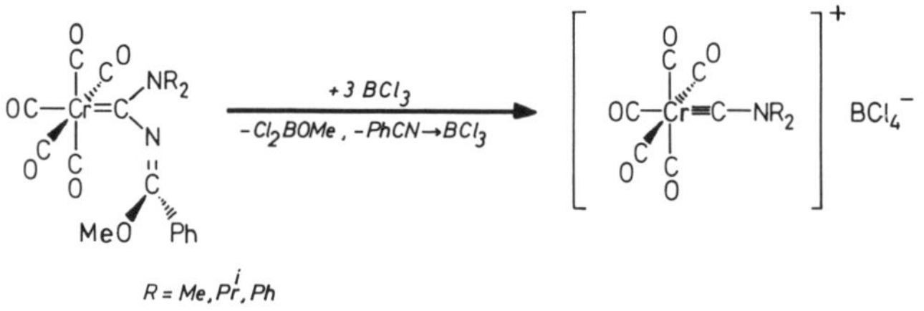

R = Me, Pr^i, Ph

Scheme 13.

Dimethylaminocarbyne complexes can also be obtained by chloride abstraction from $(CO)_5Cr[C(Cl)NMe_2]$ with silver salts or BCl_3 (Scheme 14) (70). The chlorocarbene complex can be prepared from $Na_2[Cr(CO)_5]$ and $[Me_2NCCl_2]^+Cl^-$ (71).

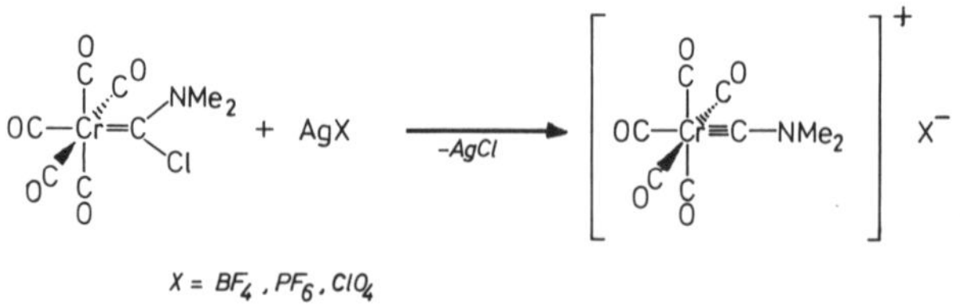

X = BF_4 , PF_6 , ClO_4

Scheme 14.

It was suggested that the formation of $(Ph_3Sn)(CO)_4M \equiv C\text{-}NMe_2$ (M = Cr, W) from $[Et_4N]_2[(Ph_3Sn)_2M(CO)_4]$ and $[Me_2NCCl_2]Cl$ also took place via a chlorocarbene complex intermediate, $(Ph_3Sn)_2(CO)_4M[C(Cl)NMe_2]$. The triphenyltin group functions as the chloride acceptor (72).

2.1.2 Rearrangement of Carbene Complexes

As already mentioned, cationic carbyne complexes, $[L_nM \equiv C\text{-}R]^+$, add any one of a great number of anionic nucleophiles, Nu^-, at the carbyne carbon atom to give carbene complexes, $L_nM[C(Nu)R]$. When heated in solution, some of the amino(Nu)carbene pentacarbonyl chromium complexes prepared by this method spontaneously rearrange with C, Cr migration of the group Nu and loss of one CO ligand to give $trans\text{-}Nu(CO)_4Cr \equiv CNR_2$ (24, 25, 38, 73–76) (Scheme 15).

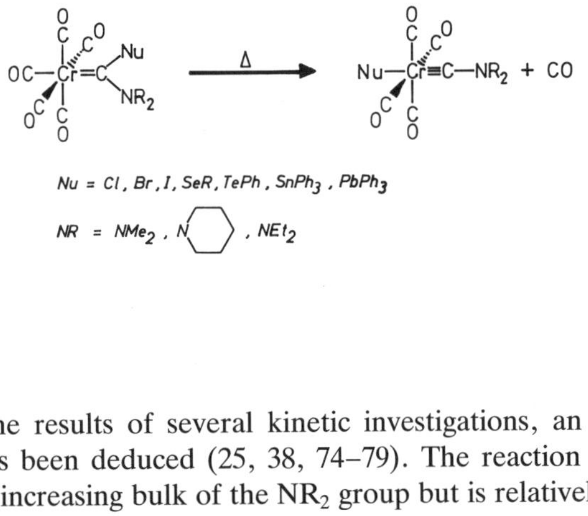

$$Nu = Cl, Br, I, SeR, TePh, SnPh_3, PbPh_3$$

$$NR = NMe_2, N\!\!\bigcirc, NEt_2$$

Scheme 15.

Based on the results of several kinetic investigations, an intramolecular mechanism has been deduced (25, 38, 74–79). The reaction rate drastically increases with increasing bulk of the NR_2 group but is relatively insensitive to variations of the migrating group Nu.

2.1.3 Reaction with Carbodiimide

Pentacarbonyl(hydroxyorganylcarbene) tungsten complexes react with dicyclohexylcarbodiimide (DCCD) via intermolecular elimination of water and substitution of one trans-CO group per two molecules of the complexes to give mixed carbene-carbyne complexes (Scheme 16) (80).

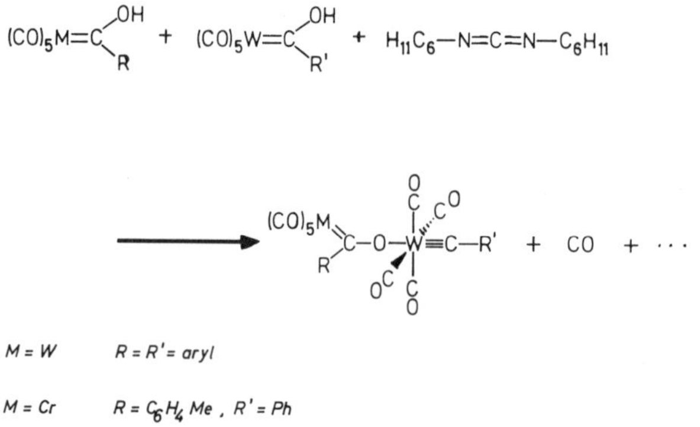

$$(CO)_5M=C\overset{OH}{\underset{R}{\diagdown}} \quad + \quad (CO)_5W=C\overset{OH}{\underset{R'}{\diagdown}} \quad + \quad H_{11}C_6-N=C=N-C_6H_{11}$$

$$\xrightarrow{\hspace{2cm}} \quad (CO)_5M\overset{}{\underset{R}{\diagdown}}C-O-W\equiv C-R' \quad + \quad CO \quad + \quad \cdots$$

M = W   R = R' = aryl

M = Cr   R = C$_6$H$_4$Me , R' = Ph

Scheme 16.

A mixed carbene-chromium/carbyne-tungsten complex is obtained from the reaction of $(CO)_5Cr[C(OH)C_6H_4CH_3]$ with $(CO)_5W[C(OH)Ph]$ and DCCD (80). In contrast, $(CO)_5Cr[C(OH)Ph]$ and DCCD afford the symmetric carbene complex anhydride $(CO)_5Cr = C(OH)\text{-}O\text{-}(OH)C = Cr(CO)_5$ (81).

2.1.4 Deprotonation of η^2-Carbene Complexes

The deprotonation of the cationic η^2-hydrido(thio)carbene complex $[HB(pz)_3](CO)_2W[\eta^2\text{-}CH(SMe)]^+$ with bases B such as NaH, K_2CO_3, NEt_3, or $NaBH_4$ gives the thiocarbyne complex $[HB(pz)_3](CO)_2W \equiv CSMe$ in generally low yield (10–20%) (82, 83). The highest yield (90%) is found when $NaSCH_2Ph$ is used as the base (83). The cationic η^2-hydrido(thio)carbene complex rapidly adds PPh_3, at room temperature, at the carbene carbon atom. Deprotonation of the $CH(PPh_3)(SMe)$ group in the resulting adduct by means of NaH affords the carbyne complex $[HB(pz)_3](CO)_2W \equiv CSMe$ in 90% yield whereas treatment with NaH of the η^2-hydrido(thio)carbene complex produces the same thio-carbyne compound in only 10% yield (84). The reaction with the secondary amines Me_2NH and Et_2NH produces air-stable aminocarbyne compounds in about 30% yield (83). It has been proposed that the reaction proceeds via an amine adduct as an intermediate (Scheme 17).

The reaction of the cationic η^2-thiocarbene complex with primary amines or NH_3 gives, in addition to small amounts of $[HB(pz)_3](CO)_2W \equiv CSMe$ (0–5%), the corresponding aminocarbyne complexes. The high air-sensitivity prevented their isolation. According to their spectra in solution, the aminocarbyne complexes are in equilibrium with their hydrideisocyanide tautomers (83).

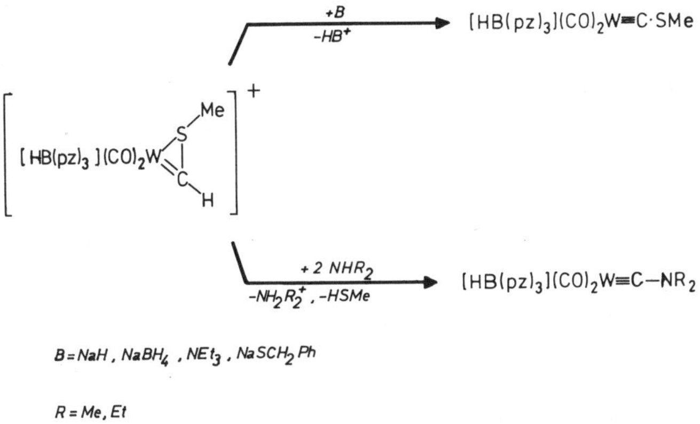

$$[HB(pz)_3](CO)_2W \equiv C \cdot SMe$$

$$\left[[HB(pz)_3](CO)_2W \overset{S-Me}{\underset{C-H}{\diagdown}} \right]^{+}$$

with:
$$\xrightarrow[-HB^+]{+B} \quad [HB(pz)_3](CO)_2W \equiv C \cdot SMe$$

$$\xrightarrow[-NH_2R_2^+ , \ -HSMe]{+ \ 2 \ NHR_2} \quad [HB(pz)_3](CO)_2W \equiv C - NR_2$$

$B = NaH \, , \ NaBH_4 \, , \ NEt_3 \, , \ NaSCH_2\,Ph$

$R = Me, Et$

Scheme 17.

Deprotonation of the carbene complex $Cl_2(CO)(PMe_3)_2W = C(H)Ph$ with the base 1-(1-cyclopentenyl)-pyrrolidine at $-78\,°C$ affords the thermally labile anionic tungsten carbyne complex $[(CH_2)_4C = N(CH_2)_4][Cl_2(CO)(PMe_3)_2W \equiv CPh]$ which on warming in solution in the presence of pyridine (py) gives $Cl(CO)(PMe_3)_2(py)W \equiv CPh$ (85). The corresponding reaction of the same carbene complex with n-butyllithium as the base followed by addition of $L = P(OMe)_3$ or Me_3CNC gives the substituted derivatives $Cl(CO)$-$(PMe_3)_2(L)W \equiv CPh$ (85).

Somewhat related to the latter dehydrochlorination reaction is the photolytic preparation of *trans*-$Me(PMe_3)_4W \equiv CMe$ from $(PMe_3)W(Me)_6$ and PMe_3 in neat PMe_3. It was proposed that the reaction proceeds via two successive α-hydrogen-methyl eliminations to afford an intermediate trimethyl carbyne complex. Subsequent methyl migration from tungsten to the carbyne carbon atom (possibly promoted by PMe_3) followed by another hydrogen-methyl elimination and addition of PMe_3 finally gives the resulting carbyne complex (86).

2.1.5 Reaction of Aryl Lithium with Halogenocarbene Complexes

Reaction of the dichlorocarbene osmium complex $Cl_2(CO)(PPh_3)_2Os = CCl_2$ with two equivalents of aryl lithium reagents, $Li[C_6H_4R\text{-}p]$ (R = H, Me, NMe$_2$) yielded trigonal-bipyramidal arylcarbyne complexes (Scheme 18) (5, 87–89).

A reduction/substitution reaction sequence was proposed very likely involving pentacoordinated $Cl(CO)(PPh_3)_2Os \equiv C\text{-}Cl$ as an intermediate (5). When other lithium reagents such as LiMe were used, product complexes, although

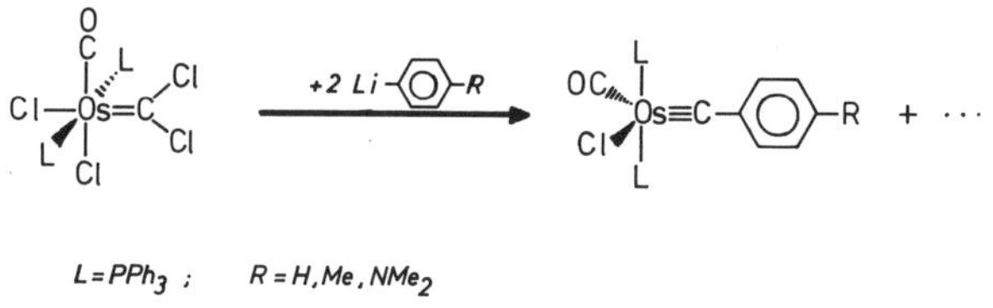

L = PPh₃ ;   R = H, Me, NMe₂

Scheme 18.

observable at low temperature, could not be isolated (5). The same is true for the product of the comparable reaction of $Cl_3(PPh_3)_2Ir=CCl_2$ with $Li[C_6H_4Me-p]$ (5).

2.2 Preparation from $L_nM-C(=O)R$ Complexes

Alkoxycarbene complexes are synthesized from acylmetallates and alkylating agents. Formal abstraction of oxide, O^{2-}, from the acyl ligand therefore seems to be a more direct route to carbyne complexes. The feasibility of this approach had already been demonstrated in 1974 (Scheme 19) (90).

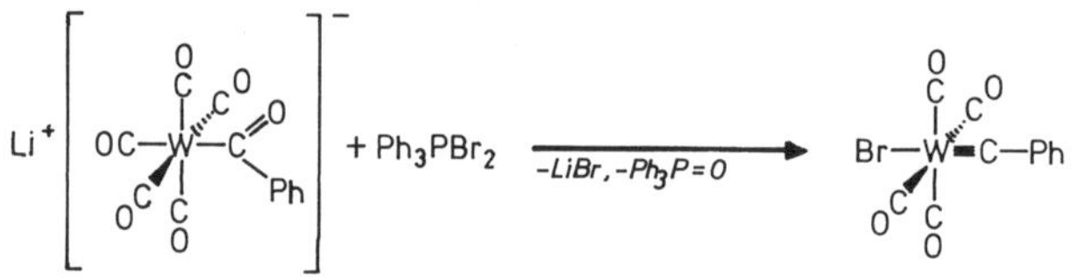

Scheme 19.

Substituting $Ph_3P(CN)Br$ for Ph_3PBr_2 also affords the trans-bromo(carbyne) complex. The corresponding trans-cyano(carbyne) complex could not be detected. For both reactions, the carbene complex $(CO)_5W[C(OPPh_3Br)Ph]$ was postulated as an intermediate (90).

Other carbyne complexes can also be prepared via this approach, e.g.

(a) $trans$-$Cl(CO)_4W\equiv CNEt_2$ from $(CO)_5W[C(O)NEt_2]^-$ and SO_2Cl_2 (91) or

(b) $trans$-$X(CO)_4M\equiv CR$ (M = Cr, Mo W; X = Cl, Br; R = alkyl, aryl) from $(CO)_5M[C(O)R]^-$ and phosgene $(COCl_2)$ or oxalyl halides $(C_2O_2X_2, X =$ Cl, Br). When the reaction is performed in the presence of nitrogen donor groups, L, bis-substituted carbyne complexes, $trans$-$X(CO)_2L_2M\equiv CR$ [L = pyridine, bipyridine, tetramethylethylenediamine (tmeda)] are formed. In a similar reaction the trans-trifluoroacetato-substituted carbyne complex *trans*-$(CF_3COO)(CO)_4W \equiv CPh$ is formed from $(CO)_5W[C-(O)Ph]^-$, tmeda and trifluoroacetic anhydride. The mono-substituted complex *trans*-$Br(CO)_3(PPh_3)W \equiv CPh$ can be obtained from $(CO)_5W-[C(O)Ph]^-$, oxalyl bromide and one equivalent of PPh_3 (92–94). However, the reaction of oxalyl bromide with the carbyne acetyltungstate $[Br(CO)_2(PPh_3)[Me(O)C]W \equiv CPh]^-$ and PPh_3 does not afford a bis-carbyne complex but rather, via carbyne-carbyne ligand coupling the alkyne complex $Br_2(CO)(PPh_3)_2W[MeC \equiv CPh]$ (Scheme 20) (95).

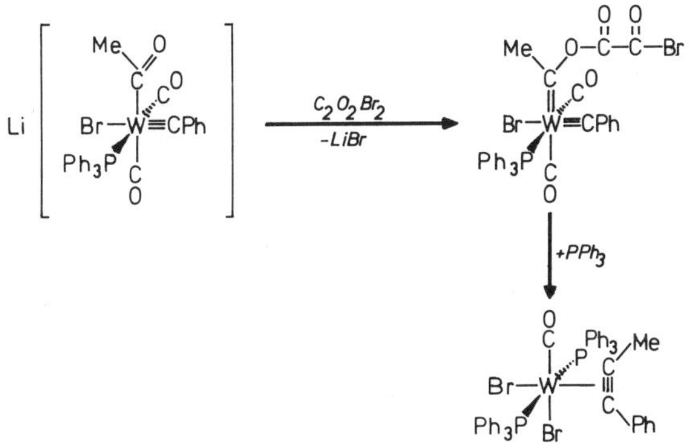

Scheme 20.

2.3 Preparation from Vinyl Complexes

Vinyl ligands may also be transformed into carbyne ligands. In benzene solution and also in the solid state at room temperature the vinyl complex $(\eta^5-C_5H_5)[P(OMe)_3]_3Mo[CH = C(H)Bu^t]$ (prepared from $[(\eta^5-C_5H_5)-[P(OMe)_3]_2Mo(HC \equiv CBu^t)]^+BF_4^-$ and $K[BH_4]$ at $-78\,°C$ in the presence of an excess of $P(OMe)_3$) rearranges to form quantitatively the carbyne complex $(\eta^5-C_5H_5)[P(OMe)_3]_2W \equiv CH_2Bu^t$ with loss of one $P(OMe)_3$ ligand (Scheme 21) (96).

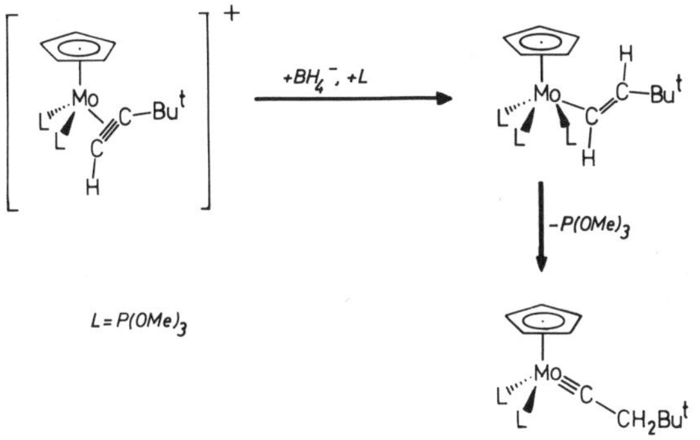

$L = P(OMe)_3$

Scheme 21.

The rearrangement reaction is suppressed by the presence of free trimethyl phosphite therefore it was suggested that a vacant coordination site is required for the H-shift to occur (96).

The thermolysis of other σ-vinyl complexes $(\eta^5\text{-}C_nH_m)[P(OMe)_3]_3M[CH = C(H)Bu^t]$ (M = Mo, $C_nH_m = C_9H_7$; M = W, $C_nH_m = C_5H_5$, C_9H_7) also leads to formation of the corresponding carbyne complexes (97). The analogous carbyne complex is obtained when $[(\eta^5\text{-}C_5H_5)[P(OMe)_3]_2Mo[HC \equiv CPr^i)]^+$ is treated with $NaBH_4/P(OMe)_3$ at room temperature, however, the similar reaction between the same cation and $K[HBBu^s_3]$ in the presence of $P(OMe)_3$ at $-78\,^\circ C$ affords an η^3-allyl complex (97).

2.4 Preparation from Metallacyclopropenes

Closely related to the preparation from vinyl complexes is that from metallacyclopropenes. The reaction of $[(\eta^5\text{-}C_9H_7)[P(OMe)_3]_2Mo(HC \equiv CSiMe_3)]^+$ BF_4^- with $K[HBBu^s_3]$ affords at low temperature a metallacyclopropene which, in toluene rearranges over several hours at room temperature by means of a 1,2-trimethylsilyl shift to form a carbyne complex in quantitative yield (Scheme 22) (97, 98).

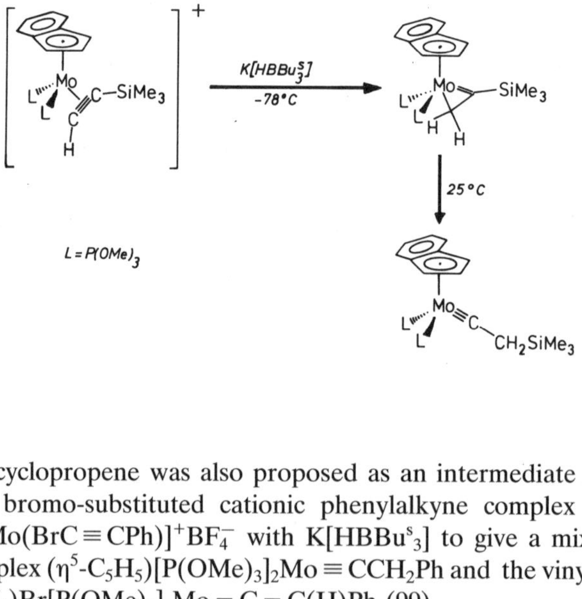

Scheme 22.

A metallacyclopropene was also proposed as an intermediate in the reaction of the bromo-substituted cationic phenylalkyne complex $[(\eta^5\text{-}C_5H_5)\text{-}[P(OMe)_3]_2Mo(BrC \equiv CPh)]^+BF_4^-$ with $K[HBBu^s_3]$ to give a mixture of the carbyne complex $(\eta^5\text{-}C_5H_5)[P(OMe)_3]_2Mo \equiv CCH_2Ph$ and the vinylidene complex $(\eta^5\text{-}C_5H_5)Br[P(OMe)_3]_2Mo = C = C(H)Ph$ (99).

3 Syntheses from Non-Carbyne Metal Complexes by Transformation of a Transition Metal Bonded sp Carbon Atom into a sp-Carbyne Carbon Atom

3.1 Preparation from $L_nM = C = X$ (X = S, Te) Complexes

By addition of electrophiles to the sulfur atom of electron-rich terminal CS ligands, thiocarbonyl groups can be transformed into thiocarbyne ligands. The NBu_4^+ salt of *trans*-$[I(CO)_4W = C = S]^-$ reacts with methyl fluorosulfonate, triethyloxonium tetrafluoroborate, acetic anhydride and trifluoroacetic anhydride, respectively, to give the corresponding neutral S-alkylated and S-acylated thiocarbyne complexes (Scheme 23) (100, 101).

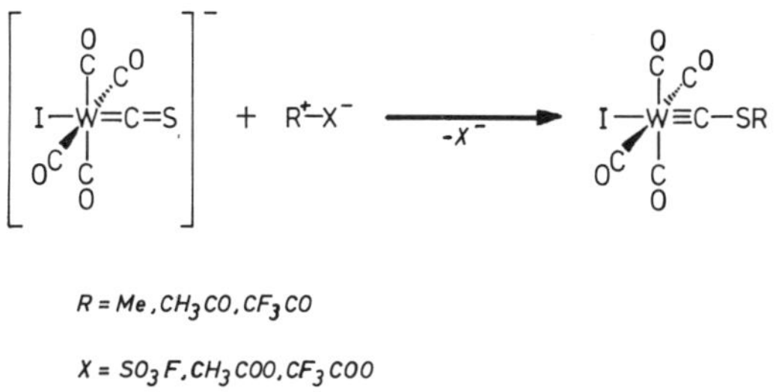

$$R = Me, CH_3CO, CF_3CO$$

$$X = SO_3F, CH_3COO, CF_3COO$$

Scheme 23.

The addition of the trifluoroacetyl group is reversible.

Cationic thiocarbyne complexes are obtained from *cis*-$(CO)(dppe)_2W = C = S$ (dppe = Ph_2P-C_2H_4-PPh_2) and $MeOSO_2F$ or $[Et_3O]BF_4$. However, the proton in CF_3SO_3H does not add to the sulfur atom but rather to the metal to form $[H(CO)(dppe)_2W(CS)]CF_3SO_3$ (101). In neither case is addition of the electrophile to the oxygen atom of a carbonyl ligand observed. In contrast to the thiocarbonyl complex *cis*-$(CO)(dppe)_2W = C = S$, the carbonyl analogue *cis*-$(CO)_2(dppe)_2W$ adds electrophiles to the metal to yield heptacoordinated σ-alkyl complexes. The formation of thiocarbyne complexes from thiocarbonyl complexes requires the sulfur atom to be highly nucleophilic. This is emphasized by the fact that neither $(CO)_5W(CS)$ nor $(CO)_4(PPh_3)W(CS)$ nor $(CO)_3(dppe)W(CS)$ affords with $[Et_3O]BF_4$ the corresponding thiocarbyne complexes (100, 101). Stable neutral thiocarbyne complexes are also obtained when the cationic thiocarbonyl compound $[HB(pz)_3(CO)_2W = C = S]^-$ ($HB(pz)_3$ = hydridotris(pyrazolyl)borate) is treated with MeI, EtI or chloro-2,4-dinitrobenzene. The analogous 2,4-dinitrophenylthiocarbyne complex can be isolated from the reaction of $[(\eta^5-C_5H_5)(CO)_2W = C = S]^-$ with chloro-2,4-dinitrobenzene. However, the related methyl- and ethylthiocarbyne complexes are less stable and can only be detected by spectroscopic means (102).

The first methyltellurocarbyne complex, $L(CO)_2W = C\text{-}TeMe$ ($L = HB(3,5-N_2C_3Me_2H)_3$) was obtained by treatment of the anionic electron-rich tellurocarbonyl complex $[L(CO)_2W = C = Te]^-$ with MeI (103). A different route constitutes the reaction of the iodo-thiocarbonyl compounds $(\eta^5-C_5H_5)(PPh_3)(I)$-$(CO)W = C = S$ (with Ph^- from LiPh) and $[HB(pz)_3](CO)_2(I)W = C = S$ (with Me^- from LiMe). With loss of iodide, phenyl- and methylthiocarbyne complexes are formed. For these reactions a charge-transfer-radical mechanism is proposed (102, 104).

3.2 Preparation from Isonitrile Complexes

Analogous to thiocarbonyl complexes, complexes with terminal isonitrile ligands can in some cases be transformed into aminocarbyne (2-azoniavinyl-idene) complexes by addition of electrophiles (H^+, R^+) to the nitrogen atom. Reaction of one molar equivalent of an acid HX (e.g. X = BF_4, HSO_4, SO_3F, $ClSO_3$, Cl, Br) with $(dppe)_2M[CNR]_2$ (M = Mo, R = Me; M = W, R = Me, Bu^t) gives monocarbyne complexes *trans*-$[(RNC)(dppe)_2M \equiv C\text{-}N(H)R]^+X$ as the kinetically controlled reaction products. These rearrange in a slower succeeding reaction to form the thermodynamically more stable hydrido-isonitrile complexes. Treatment of the hydrido-isonitrile complexes with $HBF_4 \cdot Et_2O$ gives hydrido-isonitrile-aminocarbyne complexes. However, on rapid addition of excess $HBF_4 \cdot Et_2O$ to *trans*-$(dppe)_2M[CNMe]_2$ bis(aminocarbyne) complexes are formed (105–108).

When using alkylating agents such as $MeSO_3F$, Me_2SO_4 or $[Et_3O]BF_4$ instead of protic acids, *trans*-$(dppe)_2M[CNMe]_2$ gives the cations *trans*-(MeNC)-$(dppe)_2M[CN(Me)R]^+$ (R = Me, Et) which in CH_2Cl_2 solution isomerize to form the cis-complexes. Dialkylated species cannot be obtained even in the presence of up to a 180 fold excess of the alkylating agents (105, 109).

The rhenium isonitrile complexes *trans*-$[Cl(dppe)_2Re(CNR)]$ (R = Me, Bu^t) can also be protonated by $HBF_4 \cdot Et_2O$ to give cationic aminocarbyne complexes (110, 111). By treatment of *trans*-$[Cl(dppe)_2Re(CNSiMe_3)]$ with $HBF_4 \cdot Et_2O$, finally, the unsubstituted aminocarbyne compound *trans*-$[Cl(dppe)_2Re$-$(CNH_2)]BF_4$ is accessible (112).

Analogous to the carbonyl-aminocarbyne complexes mentioned in 2.1.1, these aminocarbyne complexes are best represented as shown in Scheme 24 with form (D) having a significant weighting.

$$L_nM\equiv C-\bar{N}R_2 \longleftrightarrow L_n\bar{M}=C=\overset{+}{N}R_2$$

$$(C) \qquad\qquad (D)$$

Scheme 24.

3.3 Preparation from Vinylidene and Allenylidene Complexes

Vinylidene ligands can be converted into carbyne ligands by addition of H^+, H^- and R^-, respectively, to the β-carbon atom. The molybdenum complexes $(\eta^5\text{-}C_5H_5)[P(OMe)_3]_2(Br)Mo = C = C(H)Ph$ (I) (prepared from $(\eta^5\text{-}C_5H_5)$-$[P(OMe)_3]_2Mo[BrC\equiv CPh]^+$ and $K[HBBu_3^s]$) rapidly adds H^- from $K[HBBu_3^s]$ and Ph^- from $LiCuPh_2$, respectively, to give the corresponding carbyne complexes (Scheme 25) (113). In the course of the reaction bromide is eliminated from the complex.

A similar vinylidene compound $(\eta^5\text{-}C_5H_5)[P(OMe)_3]_2(Br)Mo = C = C(Me)Ph$ was proposed as an intermediate in the direct synthesis of $(\eta^5\text{-}C_5H_5)$-$[P(OMe)_3]_2Mo \equiv CCMe_2Ph$ from $(\eta^5\text{-}C_5H_5)[P(OMe)_3]_2Mo[BrC \equiv CPh]^+$ and $Li_2Cu_2Me_4$ (99, 113).

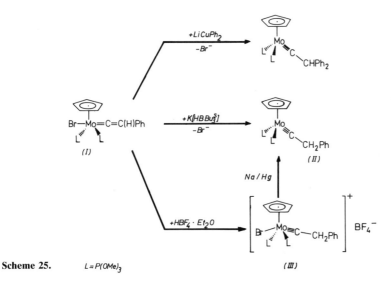

Scheme 25. $L = P(OMe)_3$

Treatment of (I) (Scheme 25) with $HBF_4 \cdot Et_2O$ gives the cationic carbyne complex (III) via H^+-addition to the β-carbon atom of (I). Reduction of (III) with sodium amalgam in tetrahydrofuran affords the carbyne complex (II) (113) (see also section 5.3).

The butyl-substituted carbyne complex $(\eta^5\text{-}C_5H_5)[P(OMe)_3]_2Mo \equiv C\text{-}CH_2Bu^t$, related to (II), can be deprotonated with $LiBu^n$ at the β-carbon atom to give $Li^+[(\eta^5\text{-}C_5H_5)[P(OMe)_3]_2Mo = C = C(H)Bu^t]^-$. Oxidation of this anionic vinylidene complex with the potential one-electron oxidants $[(\eta^5\text{-}C_5H_5)_2Fe]^+$-$BF_4^-$ or CuI in tetrahydrofuran affords a mixture of the meso-form and an RS-pair of the bis-carbyne complex $(\eta^5\text{-}C_5H_5)[P(OMe)_3]_2Mo \equiv C\text{-}C(H)(Bu^t)$-$C(H)(Bu^t)\text{-}C \equiv Mo[P(OMe)_3]_2(\eta^5\text{-}C_5H_5)$ (114).

Tungsten vinylidene complexes can also function as precursors in the synthesis of neutral as well as cationic carbyne complexes. Protonation of *mer*-$(dppe)(CO)_3W = C = C(H)Ph$ with $HBF_4 \cdot Et_2O$ affords *mer*-$[(dppe)(CO)_3W \equiv C\text{-}CH_2Ph]BF_4$, methylation with $[Me_3O]BF_4$ *mer*-$[(dppe)(CO)_3W \equiv C\text{-}C(H)$-$(Me)Ph]BF_4$ (115). The protonation is reversible, deprotonation can be achieved with 1,8-bis(dimethylamino)naphthalene (115, 116). Heating a CH_2Cl_2 solution of *mer*-$[(dppe)(CO)_3W \equiv C\text{-}CH_2Ph]BF_4$ at reflux leads to CO loss and isomerization to form $[(dppe)(CO)_2W \equiv C\text{-}CH_2Ph]BF_4$. In this compound both dppe donor atoms and the two carbonyls are now cis to the carbyne ligand. It was suggested that the trans-position was vacant although weak coordination of the counter ion BF_4^- could not be excluded. $[(dppe)(CO)_2W \equiv C\text{-}CH_2Ph]BF_4$ adds the halide from NEt_4X (even F^-!) to yield *trans*-$X(dppe)(CO)_2W \equiv C$-CH_2Ph (X = F, Cl, Br, I) (Scheme 26) (115).

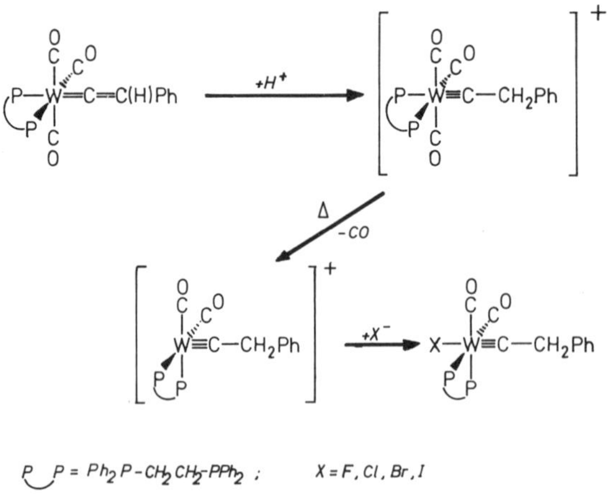

Scheme 26.

When the square-planar iridium-vinylidene complex $Cl[PPr^i_3]_2Ir = C = C-$ (H)R (R = H, Me, Ph) is heated with HBF_4 in diethyl ether, an unusual cationic compound is formed. In solution an equilibrium between the hydrido-vinylidene (E), and the carbyne form (F) is observed (Scheme 27) (117).

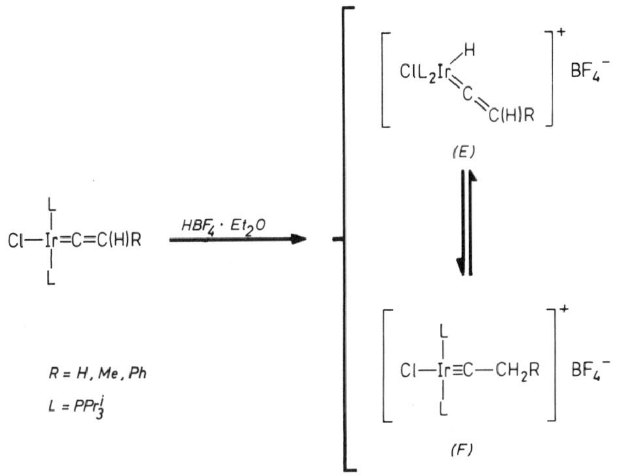

Scheme 27.

In nitromethane, the equilibrium is almost completely on the side of the carbyne form (F). Removal of the solvent shifts the equilibrium toward the side of isomer (E) and a mixture of (E) and (F) crystallizes. Addition of NaH to a suspension of (E)/(F) in benzene regenerates the neutral iridium-vinylidene complex with evolution of H_2 (117).

Vinylidene complexes are also intermediates in the preparation of carbyne complexes via double β-addition of electrophiles to acetylide ligands (see section 3.4) (118) and in the modification of the carbyne ligand of some benzylidene-carbyne complexes via consecutive reaction with $LiBu^n$ in tetrahydrofuran and electrophiles (see section 5.2) (114, 119).

Related to vinylidene ligands are allenylidene ligands. These can be used for the synthesis of vinylcarbyne complexes: protonation of the β-carbon atom in $(\eta^5\text{-}C_5H_5)(CO)_2Mn = C = C = CR_2$ (R = Ph, Bu^t) with HX (X = Cl, BF_4, CF_3COO) affords cationic vinylcarbyne complexes $[(\eta^5\text{-}C_5H_5)(CO)_2Mn \equiv C-C(H) = CR_2]^+X^-$ (120).

Iminocarbyne (2-azaallylidene) complexes, *trans*-$Br(CO)_4W \equiv C-N = CR_2$ (R = aryl), can be obtained from the reaction of the 2-azoniaallenylidene complexes $[(CO)_5W = C = N = CR_2]AlBr_4$ with freshly distilled tetrahydrofuran (26, 27). According to an X-ray analysis, the WCN fragment is strongly bent indicating that the resonance form $W = C = N = C$ contributes only to a very small extent to the overall description compared with the form $W \equiv C\text{-}N = C$. This contrasts with the observations made with aminocarbyne complexes (linear M-C-N fragment!) for which the iminocarbene (azoniavinylidene) form ((B) in Scheme 12 and (D) in Scheme 24) is of great importance.

3.4 Preparation from Acetylide Complexes

According to molecular orbital calculations (121), the gross atomic charge at the β-carbon atom of an acetylide ligand is not changed significantly by the addition of an electrophile to give a vinylidene ligand. It was therefore suggested that a second electrophile could also add to the β-carbon atom yielding a carbyne ligand. The feasibility of this approach was recently demonstrated by double β-additions of electrophiles to acetylide ligands in some tungsten complexes (118), e.g. Scheme 28.

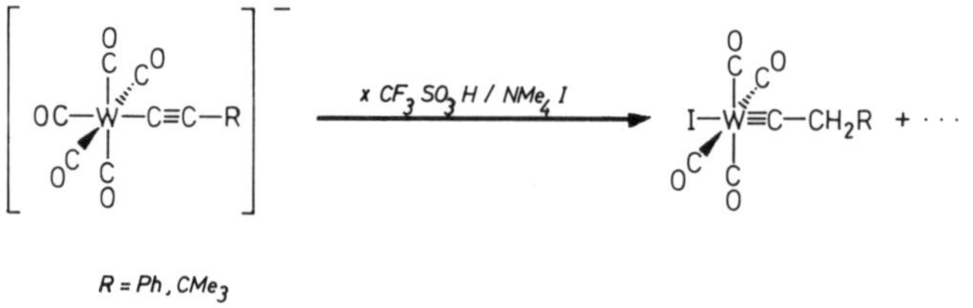

Scheme 28.

Intramolecular coupling of acetylide β-carbon atoms can lead to biscarbyne complexes. Thus the reaction of the phenylacetylide complexes $(\eta^5\text{-}C_5H_5)\text{-}(CO)_3Cr\text{-}C \equiv CPh$ with K/Na gives a binuclear biscarbyne-chromium complex (Scheme 29) (122). For the acetylide/carbyne complex conversion a complicated multi-stage redox cycle was proposed involving reduction of the triple bond with K/Na in the initial stage. Subsequent Cr-CO bond cleavage and β,β-dimerization is followed by oxidation of the reduction products in the final reaction step.

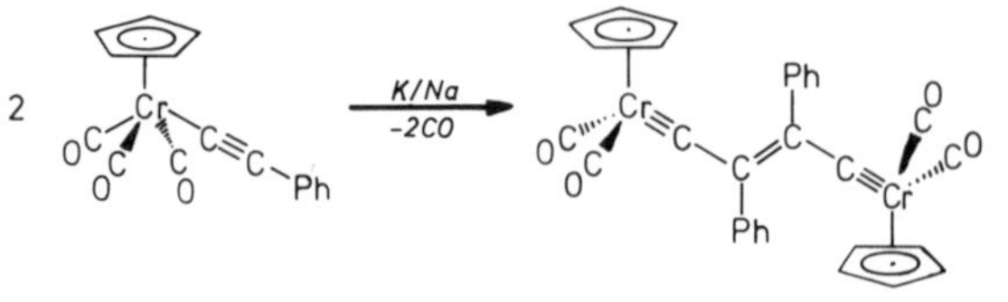

Scheme 29.

4 Miscellaneous Methods

Apart from the different synthetic routes described so far there are a few methods which do not fit into any of these categories.

Reaction of $WCl_2(PMe_3)_4$ with two equivalents of $AlMe_3$ followed by the addition of $Me_2NCH_2CH_2NMe_2$ (tmeda) gives $Cl(PMe_3)_4W \equiv CH$ (ca. 60 % isolated yield), the first complex with a terminal CH ligand (Scheme 30) (123).

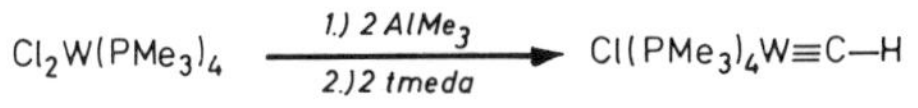

$$Cl_2W(PMe_3)_4 \quad \xrightarrow[\text{2.) 2 tmeda}]{\text{1.) 2 AlMe}_3} \quad Cl(PMe_3)_4W \equiv C-H$$

Scheme 30.

A similar complex, $Me(PMe_3)_4W \equiv CMe$, is formed by the photolysis of $(PMe_3)W(Me)_6$ in the presence of PMe_3 (86).

Another related molybdenum complex, $Me(PMe_3)_4Mo \equiv C\text{-}SiMe_3$, is obtained (among others) on thermolysis of $Br_2(PMe_3)_4Mo_2[= C(H)SiMe_3]_2$ $(M \equiv M)$ in hydrocarbon solvents (124).

When $Re_2(CO)_{10}$ is heated with $Sn(TPP)Cl_2$ (TPP = 5,10,15,20-tetraphenyl-porphyrin) at 160 °C, an unusual biscarbyne complex, $[(CO)_3Re \equiv C\text{-}]_2Sn\text{-}(TPP)$, is formed. According to the X-ray analysis, the tin atom links two $\text{-}C \equiv Re$ units (125).

The first halogenocarbyne complexes $(HBY_3)(CO)_2M \equiv C\text{-}X$ (X = Cl, Br, I; M = Mo, W; Y = 3,5-dimethyl-1-pyrazolyl) were synthesized from $[(HBY_3)\text{-}(CO)_2M]^-$ via oxidation with $Aryl\text{-}N_2^+$ or Ph_2I^+ in the presence of a source of dihalogen radicals (e.g. CH_2Cl_2, $CHBr_3$, CHI_3) (126). A mechanism involving $(HBY_3)(CO)_2M^{\cdot}$, and $CHX_2^{\cdot}$ radicals was proposed.

5 Syntheses Using Pre-existing Carbyne Complexes

In principle, carbyne complexes can be modified in various ways:
(a) by modification of the metal-ligand framework,
(b) by modification of the carbyne ligand,
(c) by oxidation or reduction of the carbyne complexes.

The most extensively used method is that of modification of the metal-ligand framework. Method (b) is far less frequently employed. At present only a few examples are known of the use of method (c). The reactions of carbyne complexes with transition metal complexes to give non-carbyne complexes, mainly metal clusters, have been reviewed elsewhere (127, 128).

5.1 Modification of the Metal-Ligand Framework

The thermal stability of carbyne complexes is strongly influenced by the ligands bonded to the metal. The carbyne ligand is a strong π-acceptor therefore the M-CO bond in carbonyl(carbyne) complexes (CO also being a π-acceptor) is labilized. Since the first reaction step in the thermal decomposition of tetracarbonyl(carbyne) complexes involves CO dissociation (129) it is to be expected that replacement of one or two CO ligands by donor groups will increase the stability of these carbyne complexes. This assumption has been confirmed by a number of experiments.

When solutions of the carbonyl(carbyne) complexes $trans$-$X(CO)_4M \equiv CR$ are treated with a slight excess of the equimolar amount of donor molecules L [such as YPh_3 (Y = P, As, Sb), $P(OPh)_3$ or PPr_3^i], the corresponding monosubstitution products are obtained (130). The formation of the monosubstituted carbyne complexes has been studied in detail (129, 131). The reaction follows a first-order rate law: $-d[\text{complex}]/dt = k[\text{complex}]$. The reaction rate, (a) is virtually independent of the electronic properties of the carbyne substituent R, (b) decreases slightly with increasing steric requirement of R, (c) increases strongly in the series trans-X = I, Br, Cl, SePh, (d) is faster for chromium complexes than for the analogous tungsten compounds by a factor of ca. 30 to 50, and (e) decreases with increasing polarity of the solvent. A dissociative mechanism has been proposed M-CO bond-rupture being rate-determining (129, 131).

Excess of L and longer reaction times (or a higher reaction temperature) generally result in the formation of di-substituted complexes $trans$-$X(CO)_2L_2M \equiv CR$ (X trans to the carbyne ligand) (130). The substitution of a second CO group, however, is not observed with chromium complexes and L = PPh$_3$ or AsPh$_3$. The arrangement of the ligands L depends on the properties of L. In general, the groups L are mutually cis but for L = P(OPh)$_3$ and M = Cr the trans-isomer is isolated (130). Irradiation with visible light of the cis-isomer (G) of X(CO)$_2$(PR$_3$)$_2$M≡CPh (M = Mo, X = Br, R = Me; M = W, X = Br, R = Ph; M = W, X = Cl, R = Me) generates the trans-isomer (H) (Scheme 31) (132).

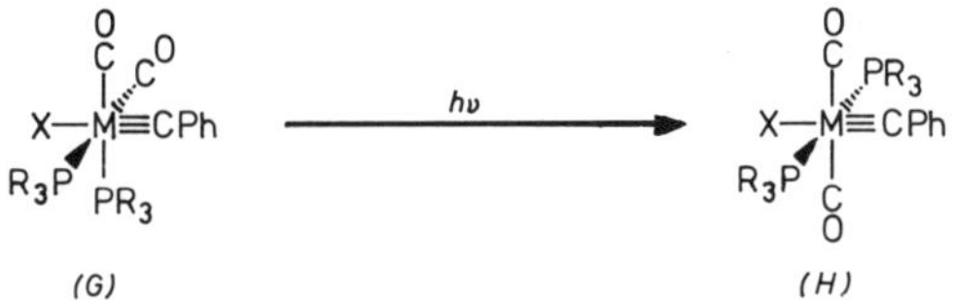

(G)   (H)

Scheme 31.

It was postulated that the isomerization proceeded via photogenerated pentacoordinate metal ketenyl complexes (132).

In contrast to other triaryl- and trialkylphosphines, PMe$_3$ can add to the carbyne carbon atom of *trans*-X(CO)$_4$Cr≡CR to give ylide-type complexes (14, 133). In some cases and when the corresponding molybdenum and tungsten complexes are employed an additional CO substitution is observed (14, 130). The product of the reaction of *trans*-Br(CO)$_4$W≡CPh with Ph$_2$As-CH$_2$-AsPh$_2$ (dam) is strongly dependent on the ratio carbyne complex/dam. For $\frac{1}{1}$ and $\frac{1}{2}$ ratios, mono- and di-substituted complexes are obtained, respectively, however, when the ratio is $\frac{2}{1}$, a dam-bridged biscarbyne complex is formed: (CO)$_2$(PhC≡)-W(μ-dam)(μ-Cl)$_2$W(≡CPh)(CO)$_2$ (134). In contrast, the related reaction of *trans*-X(CO)$_4$W≡CMe (X = Cl, Br) with dam (ratio $\frac{2}{1}$) gives via carbyne-carbyne coupling the novel 2-butyne-bridged complex (CO)$_3$W[(μ-X)(μ-dam)-(μ-MeC≡CMe)]W(CO)$_2$X (135).

Disubstituted complexes are obtained, when nitrogen-donor ligands L such as pyridine, 2,2'-bipyridine or 1,10-phenanthroline or isocyanides are employed (92–94, 130, 136, 137). Substitution of pyridine in Cl(CO)$_2$(py)$_2$W≡CR (R = Me, Ph) by PMe$_3$ gives Cl(CO)$_2$(PMe$_3$)$_2$W≡CR which is not accessible via direct substitution of CO by PMe$_3$ (94). Reaction of Cl(CO)$_2$(PMe$_3$)$_2$W≡CR (R = Ph, C$_6$H$_4$Me-*p*) (Scheme 31, isomer (G)) with neat trimethylphosphine for one week at room temperature provides Cl(CO)(PMe$_3$)$_3$W≡CR (85). Substitution of the last CO ligand can be achieved photochemically (85). The reaction of the trans-isomer of Cl(CO)$_2$(PMe$_3$)$_2$W≡CPh (Scheme 31, isomer (H)) with PMe$_3$ (1 h, 40°C, methylene chloride solution) is much faster, very likely as a consequence of the mutual trans orientation of the two π-acceptor CO ligands (132). For other routes to tetrakis(trimethylphosphine)-substituted tungsten complexes $X(PMe_3)_4W \equiv CR$ see chapter 4 (X = R = Me (86); X = Cl, R = H (123)) and chapter 5.3 (X = Br, R = Ph (85)).

Tris(trimethylphosphite)-substituted carbyne complexes $X(CO)[P(OMe_3)_3]_3$-$M \equiv CPh$ can be prepared by treatment of $Cl(CO)_4M \equiv CPh$ (M = Mo, W), $Cl(CO)_2(py)_2M \equiv CPh$ (M = Mo, W) and $Br(CO)_2(py)_2Cr \equiv CPh$, respectively, with $P(OMe)_3$. On prolonged heating of the molybdenum and tungsten complexes in neat trimethylphosphite, even the substitution of the remaining CO ligand by $P(OMe)_3$ is possible (93). The phosphite ligands in *trans*-Cl-$[P(OMe)_3]_4W \equiv CPh$ are coordinatively labile and can be substituted e.g. by $Ph_2P\text{-}CH_2CH_2\text{-}PPh_2$ to give *trans*-$Cl(dppe)_2W \equiv CPh$ (93). Reaction of halides Y^- with neutral $X(CO)_4W \equiv CR$ (R = aryl) affords labile anionic dihalogeno tricarbonyl carbyne complexes, $[mer\text{-}XY(CO)_3W \equiv CR]^-$. These are stable only in solution in the presence of excess halide (138). The same is true for NEt_4^+ $[Cl_2(CO)_2(py)W \equiv CR]^-$ (R = Me, Ph) formed by treatment of $Cl(CO)_2$-$(py)_2W \equiv CR$ with excess $NEt_4^+Cl^-$ (139). The latter anionic complexes react with maleic anhydride or fumaronitrile to give unusual anionic alkene carbyne complexes, $NEt_4^+[Cl_2(py)(CO)(alkene)W \equiv CR]^-$, which are stable and can be isolated in a pure form (139). Neutral alkene carbyne complexes, $Cl(CO)(py)_2$-$(alkene)W \equiv CR$ (Cl and CR, py and CO, and Cl and alkene being mutually trans), are obtained from $Cl(CO)_2(py)_2W \equiv CPh$ and a large excess of maleic anhydride or fumaronitrile at elevated temperatures (40–50 °C) (139).

Treatment of the carbonyl-free carbyne complex *trans*-$Br(PMe_3)_4W \equiv CPh$ with two equivalents of HBr gas provides the hydrido tungsten carbyne complex $Br_2(H)(PMe_3)_3W \equiv Ph$ in 95 % yield (85).

Heating a CH_2Cl_2 solution of $[mer\text{-}(dppe)(CO)_3W \equiv CCH_2Ph]BF_4$ with NEt_4X (X = Cl, Br, I) leads to halide substitution for one CO ligand and isomerization to give $X(dppe)(CO)_2W \equiv CCH_2Ph$ (X and CCH_2Ph now trans) (115). Since the anion $[mer\text{-}(dppe)(CO)_3W \equiv CCH_2Ph]^-$ easily loses one CO group at elevated temperature to form $[(dppe)(CO)_2W \equiv CCH_2Ph]^-$, a stepwise substitution can be carried out. On addition of L = PMe_3, acetone, water, and NEt_4X (X = F, Cl, Br, I), respectively, the complexes $[L(dppe)(CO)_2W \equiv CCH_2Ph]BF_4$ are easily prepared (115).

The type of product obtained from $[L(CO)_2M \equiv CR]^+$ (M = Mo, L = $C_5H_5^-$; M = Cr, L = C_6H_6) and PMe_3 varies strongly. The molybdenum compound $(\eta^5\text{-}C_5H_5)(CO)_2Mo \equiv CR(R = \text{-}C_6H_4Me)$ reacts with one equivalent of PMe_3 via substitution of one CO ligand (140) and with two equivalents of PMe_3 to form a ketenyl complex, $(\eta^5\text{-}C_5H_5)(CO)[PMe_3]_2Mo[C(CO)R]$ (141). In contrast, PMe_3 adds to the carbyne carbon of $[(\eta^6\text{-}C_6H_6)(CO)_2Cr \equiv CPh]^+$ to give ylide complexes, $[(\eta^6\text{-}C_6H_6)(CO)_2Cr(C(PMe_3)Ph)]^+$ (142).

The reactivity of aminocarbyne (2-azoniavinylidene) complexes toward nucleophiles has also been studied extensively. In general, anionic nucleophiles

X^- such as Cl^-, Br^-, I^-, SCN^-, CN^-, SR^-, SeR^- ... add to the carbyne carbon atom of $[(CO)_5Cr \equiv CNEt_2]^+$ to give amino(X)carbene complexes (68) (for the transformation of aminocarbene into aminocarbyne complexes see 2.1.2). However, PPh_3 does not add to the carbyne carbon but rather the substitution of a CO ligand (very likely the trans-CO group) by PPh_3 is observed yielding *trans*-$[PPh_3(CO)_4Cr \equiv CNEt_2]^+$ (143, 144) which, in contrast to $[(CO)_5Cr \equiv CNEt_2]^+$, does not afford carbene complexes on addition of halides X^-. Instead neutral carbyne complexes $X(CO)_3(PPh_3)Cr \equiv CNEt_2$ (X and $CNEt_2$ trans) are obtained (143, 144).

The related cationic molybdenum complex $[(CO)_5Mo(CNEt_2)]BF_4$ even at $-40\,°C$ easily loses one carbonyl ligand to form neutral *trans*-$(BF_4)(CO)_4Mo(CNEt_2)$. The low temperature reaction of the cationic compound with X^- (X = Br, I, SCN) yields neutral *trans*-$X(CO)_4Mo(CNEt_2)$ complexes and with X^- (X = OCN, $SeC_6H_4CF_3$) binuclear X-bridged biscarbyne complexes $(\mu\text{-}X)_2[(CO)_3Mo(CNEt_2)]_2$ (66). Similar binuclear molybdenum and tungsten complexes (X = Br, I, SCN, SPh, SePh, $SeC_6H_4F\text{-}p$, $SeC_6H_4CF_3\text{-}p$) are prepared from the corresponding neutral carbyne complexes $X(CO)_4M(CNEt_2)$ (M = Mo, W) on warming in solution (66).

Comparable to the related molybdenum complex, $[(CO)_5W(CNEt_2)]Y$ (Y = BF_4, SbF_6, $SbCl_6$) also readily loses one CO ligand giving $Y(CO)_4W(CNEt_2)$ (65). On addition of X^- (X = CN, NCO, NCS, NCSe, SPh), Y^- is rapidly displaced and *trans*-$X(CO)_4W(CNEt_2)$ is formed (65). Direct substitution of CO is observed when $[(CO)_5W(CNEt_2)]^+$ is treated with SeR^- (R = Ph, $C_6H_4F\text{-}p$, $C_6H_4CF_3\text{-}p$) (64).

The reactivity of neutral aminocarbyne complexes parallels that of the corresponding alkyl- and arylcarbyne complexes. When treated with neutral donor ligands such as phosphines, 2,2'-bipyridine (bipy) or 1,10-phenanthroline (ophen), the substitution of one or two CO ligands is observed (129, 144, 145). The resulting substitution products are thermally somewhat more stable than the corresponding substituted aryl- and alkylcarbyne complexes. This effect has been used for the preparation and isolation of anionic dihalogeno(aminocarbyne) complexes $M^+[mer\text{-}X_2(CO)_3W(CNR_2)]^-$ (M = NEt_4, X = Br, R = Hex^c; M = PPN, X = Cl, R = Hex^c, Pr^i) (146). The cis-halide in these anionic complexes is labile and can easily be displaced by neutral nucleophiles such as PPh_3 yielding once more neutral carbyne complexes (146). The displacement of two CO ligands by two halide ions in $X(CO)_4W \equiv CNR_2$ to yield dianionic carbyne complexes could not be observed (146). However, the reaction of *trans*-$(p\text{-}MeC_6H_4S)(CO)_4W(CNEt_2)$ with $NEt_4[SCN]$ in excess afforded a dianionic complex, $[NEt_4]_2[(SCN)_3(CO)_2W(CNEt_2)]$ (147). Another aminocarbyne complex anion, $[Br_2(CO)_3W(CNHex_2^c)]^-$, was obtained from $Br(CO)_4W(CNHex_2^c)$ and an electron-rich olefin (148). The halide ligand in disubstituted aminocarbyne complexes of the type $Br(CO)_2L_2W(CNEt_2)$ (L_2 = bipy, ophen) turned out to be easily replaceable. The reactions with $NaAsPh_2$, KCN, and $[(CO)_5-M-EPh_2]^-$ (M = Cr, Mo, W; E = P, As, Sb) gave *trans*-$(Ph_2As)(CO)_2(bipy)-W(CNEt_2)$ (149), $(CN)(CO)_2L_2W(CNEt_2)$ (145), and $(CO)_5M(\mu-EPh_2)-(CO)_2L_2W(CNEt_2)$ (147), respectively. The latter two types of complexes were found to exhibit a dynamic behaviour. A symmetrical intermediate was proposed in the dynamic process, based upon the fluctuation of the chelate ligand L_2 between cis/trans positions to the carbyne ligand (142). When $X(CO)_2L_2-W(CNEt_2)$ (X = Br, I) was treated with *cis*-$[Mo(CO)_4(PPh_2K)_2]$, the L_2 and X ligands were displaced and a dinuclear anionic carbyne complex, $K[(CO)_4-Mo(\mu-PPh_2)_2W(CO)_2(CNEt_2)]^-$, was obtained (Scheme 32) which adds CO_2 to the $W \equiv C$ bond to give a metallacyclic complex containing the fragment $\overline{W = C(NEt_2)C(O)O}$ (151).

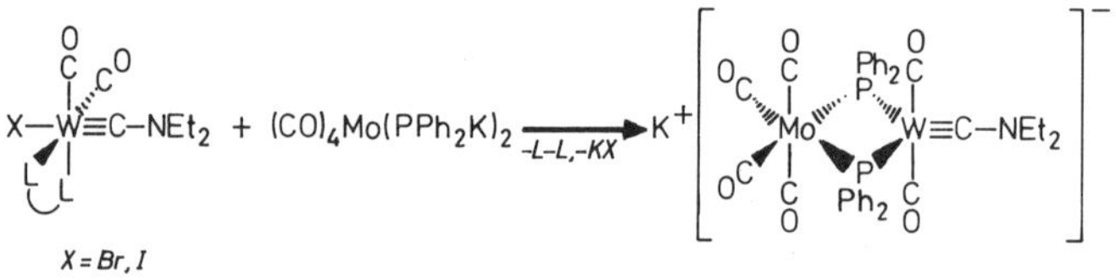

$X = Br, I$

$L \smile L = 2,2'-bipy, 1,10-ophen$

Scheme 32.

The reaction of *trans*-$I(CO)_2L_2W(CNEt_2)$ with $Na_2[S_2C_2(CN)_2]$ afforded, via substitution of the halide ligand and elimination of the chelating ligand L_2 (bipy, ophen), a pentacoordinated anionic carbyne complex, $[((CN)_2C_2S_2)-(CO)_2W(CNEt_2)]^-$. The complex was isolated as the NEt_4^+ salt (152). The halide ligands in *trans*-$X(CO)_4M \equiv CR$ (R = alkyl, aryl) are also substitution-labile and are easily displaced by other halides. This is of advantage (and has extensively been used (13, 18, 29, 33)) for the preparation of trans-iodide substituted complexes. Although these are, in the series X = Cl, Br, I, the thermally most stable ones they are less easily synthesized in good yield from carbene complexes and Lewis acids because of some difficulties in handling the corresponding Lewis acids. The halide exchange (Scheme 33) is an equilibrium reaction. Thermodynamic factors favor the complex containing the heavier halide. By proper choice of the solvent, the equilibrium can be reversed (138). Kinetic investigations show that the substitution reaction follows a second-order rate law, first order in the concentrations of the complex and the halide. The influence of the leaving group on the rate is very small, coordinated iodide is substituted more than three times faster by chloride than by bromide. A mechanism involving a nucleophilic, frontier-orbital controlled attack of the halide ion at the carbyne complex in the rate-determining step was proposed (153).

$$X,Y = Cl, Br, I$$

Scheme 33.

The reaction of *trans*-$Br(CO)_4M \equiv CR$ (M = Cr, W; R = Ph, C_6H_4Me-*p*) with K[OH], Na[OEt], and LiPh, respectively, did not yield the corresponding trans-substituted complexes but rather organic products (RCH_2-CO-OH, RCH_2-CO-OEt, RCH_2-CO-Ph) were isolated (154).

A series of carbyne complexes which are difficult to prepare or are even inaccessible by other routes have been synthesized via displacement of the BF_4^- ligand (BF_4^- being an excellent leaving group) by Y^- in labile *trans*-$(BF_4)(CO)_3LM \equiv CR$ (L = CO; M = W; R = Me, Ph; Y^- = PPh_3, SCN^-, CN^- (42), $AsPh_3$, Bu^tNC (43); L = PMe_3; M = Cr; R = Me; Y^- = PPh_3 (42), H_2O (45)). Other leaving groups which have been used include $(CO)_5M[C(O)R]$ (M = Cr, W; R = Me, aryl) (80, 81).

Carbyne complexes with a metal-metal bond can be prepared via substitution of the halide ligand in e.g. *trans*-$X(CO)_4 M \equiv CR$ (X = Cl, Br; M = Cr, Mo, W; R = Ph, $SiPh_3$, menthyl) by carbonyl metallates such as $[(CO)_5M]^-$ (M = Mn, Re) (Scheme 35) (7, 29, 155), $[(\eta^5-C_5H_5)(CO)_3M]^-$ (M = Mo, W) (13), and $[(CO)_4Co]^-$ (156). Such an exchange significantly increases the stability of the carbyne complexes.

A carbyne complex containing a Cr-Br-Cr fragment, $(CO)_5Cr$-Br-$Cr(CO)_4$- ($\equiv C$-C_3H_5), is obtained on thermolysis of *trans*-$Br(CO)_4Cr \equiv C$-C_3H_5 in dichloromethane at $-30\,°C$ (11). The reaction of *trans*-$X(CO)_4M \equiv CR$ (X = Cl, Br, I; M = Cr, W; R = Me, Ph) with $(CO)_5Cr(CH_2Cl_2)$, generated by photolysis of $Cr(CO)_6$ in dichloromethane, gives similar halide-bridged binuclear Cr-X-M complexes (Scheme 34) (157) which are readily cleaved, even by weak donors L (L = Et_2O, tetrahydrofuran, PPh_3, Br^-), to form the corresponding carbyne complexes and the pentacarbonyl chromium derivative of the donor L, $(CO)_5Cr$-L (157).

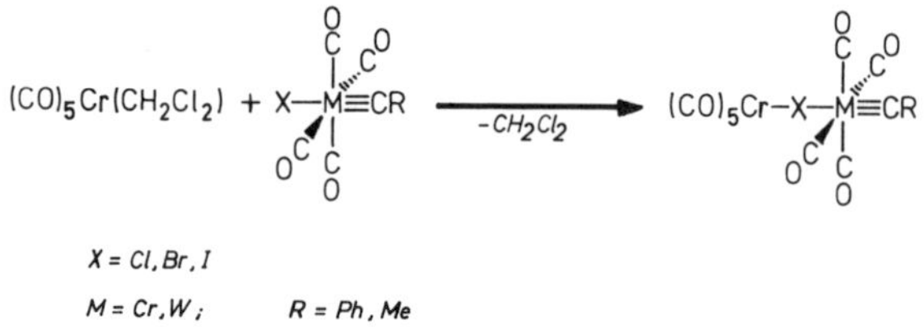

(CO)5Cr(CH2Cl2) + X—M≡CR
-CH2Cl2
(CO)5Cr—X—M≡CR
X = Cl, Br, I
M = Cr, W;
R = Ph, Me

Scheme 34.

When *trans*-$Br(CO)_4W{\equiv}CR$ complexes are treated with lithium or sodium cyclopentadienide not only the displacement of the halide but an additional elimination of two carbonyl ligands is observed and $(\eta^5\text{-}C_5H_5)(CO)_2W{\equiv}CR$ complexes (R = aryl, NEt_2 (158, 159), $(\eta^5\text{-}C_5H_4)Fe(\eta^5C_5H_5)$ (159), $SiPh_3$, $SiPh_2Me$, (28, 29)) are formed (Scheme 35).

+Na[(CO)5Mn]
-NaBr
(CO)5Mn—W≡CPh
Br—W≡CPh
+LiC5H5
-LiBr, -2CO
W≡CPh

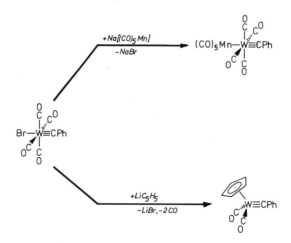

Scheme 35.

The reaction of $Br(CO)_4W{\equiv}CSiPh_3$ with lithium indenide yields an analogous complex. However, $[(\eta^5\text{-}C_5H_5)(CO)_2Mn{\equiv}CMe]^+$ adds $C_5H_5^-$ at the carbyne carbon atom to give a carbene complex (51). Similarly, treatment of *trans*-$Br(CO)_4W{\equiv}CR$ (a) with $K[R'B(pz)_3]$ (R' = H, pz (pyrazolyl)) affords $(R'B(pz)_3)(CO)_2W{\equiv}CR$ (R = Me, C_6H_4Me-*p*) (160, 161) and (b) with $Na_2[7,8\text{-}C_2B_9H_9Me_2]$ followed by metathesis with $[N(PPh_3)_2]Cl$ gives $[N(PPh_3)_2][(\eta\text{-}1,2\text{-}C_2B_9H_9Me_2)(CO)_2W{\equiv}CR]$ (R = C_6H_4Me-*p*) (162, 163).

Some carbyne complexes can be protonated at the metal, e.g. the cationic carbyne hydride complex $(\eta^5\text{-}C_5H_5)[P(OMe)_3]_2(H)Mo{\equiv}C\text{-}CH_2Bu^t]BF_4$ is formed on protonation of $(\eta^5\text{-}C_5H_5)[P(OMe)_3]_2Mo{\equiv}C\text{-}CH_2Bu^t$ with HBF_4

(164). Protonation at the metal is also observed in the reaction of $Cl(PMe_3)_4W \equiv CH$ with HCl and of $Cl(dppe)_2W \equiv CR$ (R = H, CMe_3) with HCl or CF_3SO_3H (165, 166).

A carbene complex formed on protonation of the carbyne carbon atom of $(\eta^5\text{-}C_5H_5)(CO)_2W \equiv CR$ (R = Me, $C_6H_4Me\text{-}p$) was proposed as an intermediate in the reaction with $HBF_4 \cdot Et_2O$ or with HI. Different noncarbyne complexes, depending on R and the acid used, were isolated (167).

Protonation of the closely related salts $[N(PPh_3)_2][(\eta\text{-}1,2\text{-}C_2B_9H_9Y_2)(CO)_2\text{-}W \equiv CR$ (R = $-C_6H_4Me]$ also affords structurally different products: if Y = Me a carbyne complex, $(\eta\text{-}C_2B_9H_{10}Me_2)(CO)_2W \equiv CR$, is formed, however, if Y = H a hydrogen- and alkyne-bridged binuclear complex is obtained (168). The zerovalent osmium carbyne complex $Cl(CO)(PPh_3)_2Os \equiv CR$ (R = $C_6H_4NMe_2$) reacts with molecular oxygen giving a $\frac{1}{1}$ adduct which was formulated as a divalent, octahedral complex retaining the unchanged carbyne ligand, and with a dihapto-peroxycarbonyl ligand, $Cl(PPh_3)_2(O_2CO)Os \equiv CR$. The reaction with HCl liberated CO_2 and gave $[Cl_2(H_2O)(PPh_3)_2Os \equiv CR]^+$ the H_2O ligand of which could be replaced by SCN^- and MeC_6H_4NC (169).

5.2 Modification of the Carbyne Ligand

As shown in chapter 5.1, the metal-ligand framework of carbyne complexes can be altered in various ways. In contrast, the possibilities for the modification of the carbyne ligand are much more restricted. Not surprisingly, there are only a limited number of reports about the synthesis of carbyne complexes via this route. The reactions reported so far fall into two categories:
(a) modifications within the group R of $L_nM \equiv C\text{-}R$ and
(b) replacement of R by R'.

The ethynylcarbyne complexes $trans\text{-}X(CO)_4W \equiv C\text{-}C \equiv CPh$ (X = Cl, Br, I) react with dimethylamine even at $-25\,°C$ with addition of the amine to the carbon-carbon triple bond to give $trans\text{-}X(CO)_4W \equiv C\text{-}CH = C(NMe_2)Ph$ (19).

The complexes $(\eta^5\text{-}C_5H_5)[P(OMe)_3]_2M \equiv CCH_2Bu^t$ (M = Mo, W) can be deprotonated with Bu^nLi to give the anionic species $Li[(\eta^5\text{-}C_5H_5)[P(OMe)_3]_2\text{-}M(CCHBu^t)]$ which (a) on quenching with D_2O afford a mixture of $(\eta^5\text{-}C_5H_5)[P(OMe)_3]_2M \equiv CCHDBu^t$ and $(\eta^5\text{-}C_5H_5)[P(OMe)_3]_2M \equiv CCD_2Bu^t$ (119) and (b) add a range of electrophiles R^+ (R e.g. Me, CH_2CH_2OH, $COBu^t$, $SiMe_3$, SMe, $CH_2CH = CH_2$) to give the substituted carbyne complexes $(\eta^5\text{-}C_5H_5)[P(OMe)_3]_2Mo \equiv CCH(R)Bu^t$ (114).

The trimethylsilyl-substituted carbyne complexes $(\eta^5\text{-}C_9H_7)[P(OMe)_3]_2\text{-}Mo \equiv C\text{-}CH(R)SiMe_3$ (R = H, Ph) are desilylated in NaF/aqueous MeCN to form $(\eta^5\text{-}C_9H_7)[P(OMe)_3]_2Mo \equiv CCH_2R$ (97, 98).

Nucleophilic displacement of the carbyne substituent R has been observed in the reaction of $L(CO)_2Mo \equiv CR$ [R = Cl; L = $HB(3,5-N_2C_3Me_2H)_3$] with ER^- (SMe, SPh, $SC_6H_4NO_2$-p, SePh) to give $L(CO)_2Mo \equiv C$-ER (103).

Treatment of $L(CO)_2W \equiv CSMe$ (L = $HB(3,5-N_2C_3Me_2H)_3$) with excess PEt_3 affords $[L(CO)_2W \equiv C$-$PEt_3]^+$, with PMe_3, however, $[L(CO)_2W = C$-$(PMe_3)_2]^+$ is obtained which on addition of MeI gives $[L(CO)_2W \equiv C$-$PMe_3]^+$ (170).

5.3 Oxidation and Reduction of Carbyne Complexes

Oxidation and reduction offer another possibility for the modification of carbyne complexes although this means have not been used very often until now. Examples include

(a) the reduction of $[(\eta^5$-$C_5H_5)[P(OMe)_3]_2BrMo \equiv C$-$CH_2Ph]BF_4$ with Na/Hg (113) or Mg/Hg (99),

(b) the reductions of $Br_3(dme)W \equiv CPh$ (dme = dimethoxyethane) and Br_3-$(PMe_3)_3Mo \equiv CPh$, respectively, with zinc in acetonitrile in the presence of PMe_3 to give $Br(PMe_3)_4M \equiv CPh$ (M = Mo, W) (85), and

(c) the oxidation of *trans*-$Br(CO)_4M \equiv CR$ (M = Mo, W; R = Me, Ph, CH_2CMe_3) with bromine in dimethoxyethane (dme) to give $Br_3(dme)M \equiv CR$ (171).

However, the reaction of *trans*-$Br(CO)_4Cr \equiv CPh$ with Cl_2, Br_2, cerium(IV) salts, and manganese(III) compounds results in the complete degradation of the complex. Benzotrihalides or benzoic acid derivatives and, in some cases, also dibenzyl are isolated as products of the carbyne ligand (172).

6 References

(1) E. O. Fischer, G. Kreis, C. G. Kreiter, J. Müller, G. Huttner, H. Lorenz, Angew. Chem. **85** (1973) 618–620; Angew. Chem. Int. Ed. Engl. **12** (1973) 564–565.

(2) E. O. Fischer, Angew. Chem. **86** (1974) 651–663; Adv. Organomet. Chem. **14** (1976) 1–32.

(3) E. O. Fischer, U. Schubert, J. Organomet. Chem. **100** (1975) 59–81.

(4) U. Schubert, in F. R. Hartley, S. Patai (Eds.): *The Chemistry of the Metal-Carbon Bond,* John Wiley and Sons, Chichester 1982, pp. 233–243.

(5) M. A. Gallop, W. R. Roper, Adv. Organomet. Chem. **25** (1986) 121–198.

(6) E. O. Fischer, G. Kreis, Chem. Ber. **109** (1976) 1673–1683.

(7) S. Fontana, O. Orama, E. O. Fischer, U. Schubert, F. R. Kreissl. J. Organomet. Chem. **149** (1978) C57–C62.

(8) E. O. Fischer, Nguyen Quy Dao, W. R. Wagner, Angew. Chem. **90** (1978) 51; Angew. Chem. Int. Ed. Engl. **17** (1978) 50–51.

(9) Nguyen Quy Dao, E. O. Fischer, W. R. Wagner, D. Neugebauer, Chem. Ber. **112** (1979) 2552–2564.

(10) Nguyen Quy Dao, M. Jouan, G. P. Fonseca, N. Hoa Tran Huy, E. O. Fischer, J. Organomet. Chem. **287** (1985) 215–219.

(11) E. O. Fischer, N. Hoa Tran-Huy, D. Neugebauer, J. Organomet. Chem. **229** (1982) 169–177.

(12) E. O. Fischer, A. Schwanzer, H. Fischer, D. Neugebauer, G. Huttner, Chem. Ber. **110** (1977) 53–66.

(13) E. O. Fischer, T. L. Lindner, F. R. Kreißl, P. Braunstein, Chem. Ber. **110** (1977) 3139–3148.

(14) F. R. Kreißl, W. Uedelhofen, G. Kreis, Chem. Ber. **111** (1978) 3283–3293.

(15) E. O. Fischer, U. Schubert, H. Fischer, Inorg. Synth. **19** (1979) 172–174.

(16) H. Fischer, F. Seitz, Organomet. Synth. **3** (1986) 209–215.

(17) Nguyen Quy Dao, E. O. Fischer, C. Kappenstein, Nouveau J. Chimie **4** (1980) 85–94.

(18) E. O. Fischer, W. R. Wagner, F. R. Kreißl, D. Neugebauer, Chem. Ber. **112** (1979) 1320–1328.

(19) E. O. Fischer, H. J. Kalder, F. H. Köhler, J. Organomet. Chem. **81** (1974) C23–C27.

(20) E. O. Fischer, T. Selmayr, Z. Naturforsch. **32b** (1977) 105–107.

(21) E. O. Fischer, G. Kreis, F. R. Kreißl, W. Kalbfus, E. Winkler, J. Organomet. Chem. **65** (1974) C53–C56.

(22) E. O. Fischer, G. Huttner, W. Kleine, A. Frank, Angew. Chem. **87** (1975) 781; Angew. Chem. Int. Ed. Engl. **14** (1975) 760.

(23) E. O. Fischer, W. Kleine, G. Kreis, F. R. Kreißl, Chem. Ber. **111** (1978) 3542–3551.

(24) E. O. Fischer, W. Kleine, F. R. Kreißl, H. Fischer, P. Friedrich, G. Huttner, J. Organomet. Chem. **128** (1977) C49–C53.

(25) H. Fischer, A. Motsch, R. Märkl, K. Ackermann, Organometallics **4** (1985) 726–735.

(26) H. Fischer, F. Seitz, J. Riede, J. Chem. Soc., Chem. Commun. (1985) 537–539.

(27) H. Fischer, F. Seitz, J. Riede, Chem. Ber. **119** (1986) 2080–2093.

(28) E. O. Fischer, H. Hollfelder, P. Friedrich, F. R. Kreißl, G. Huttner, Angew. Chem. **89** (1977) 416–417; Angew. Chem. Int. Ed. Engl. **16** (1977) 401–402.

(29) E. O. Fischer, H. Hollfelder, F. R. Kreißl, Chem. Ber. **112** (1979) 2177–2189.

(30) E. O. Fischer, V. N. Postnov, F. R. Kreißl, J. Organomet. Chem. **127** (1977) C19–C21.

(31) E. O. Fischer, F. J. Gammel, D. Neugebauer, Chem. Ber. **113** (1980) 1010–1019.

(32) E. O. Fischer, M. Schluge, J. O. Besenhard, Angew. Chem. **88** (1976) 719–720; Angew. Chem. Int. Ed. Engl. **15** (1976) 683–684.

(33) E. O. Fischer, M. Schluge, J. O. Besenhard, P. Friedrich, G. Huttner, F. R. Kreißl, Chem. Ber. **111** (1978) 3530–3541.

(34) E. O. Fischer, F. J. Gammel, J. O. Besenhard, A. Frank, D. Neugebauer, J. Organomet. Chem. **191** (1980) 261–282.

(35) E. O. Fischer, W. Röll, N. Hoa Tran Huy, K. Ackermann, Chem. Ber. **115** (1982) 2951–2964.

(36) Nguyen Quy Dao, H. Fevrier, M. Jouan, E. O. Fischer, W. Röll, J. Organomet. Chem. **275** (1984) 191–207.

(37) K. Weiß, E. O. Fischer, Chem. Ber. **109** (1976) 1868–1886.

(38) E. O. Fischer, H. Fischer, U. Schubert, R. B. A. Pardy, Angew. Chem. **91** (1979) 929–930; Angew. Chem. Int. Ed. Engl. **18** (1979) 871.

(39) A. Schwanzer, Dissertation, Technische Universität München, 1976.

(40) E. O. Fischer, T. Selmayr, F. R. Kreißl, U. Schubert, Chem. Ber. **110** (1977) 2574–2583.

(41) E. O. Fischer, S. Walz, W. R. Wagner, J. Organomet. Chem. **134** (1977) C37–C39.

(42) E. O. Fischer, S. Walz, A. Ruhs, F. R. Kreißl, Chem. Ber. **111** (1978) 2765–2773.

(43) E. O. Fischer, F. J. Gammel, Z. Naturforsch. **34b** (1979) 1183–1185.

(44) E. O. Fischer, K. Richter, Chem. Ber. **109** (1976) 2547–2557.

(45) K. Richter, E. O. Fischer, C. G. Kreiter, J. Organomet. Chem. **122** (1976) 187–196.

(46) E. O. Fischer, K. Richter, Angew. Chem. **87** (1975) 359–360; Angew. Chem. Int. Ed. Engl. **14** (1975) 345–346.

(47) E. O. Fischer, K. Richter, Chem. Ber. **109** (1976) 3079–3088.

(48) E. O. Fischer, P. Rustemeyer, J. Organomet. Chem. **225** (1982) 265–277.

(49) E. O. Fischer, R. L. Clough, G. Besl, F. R. Kreißl, Angew. Chem. **88** (1976) 584–585; Angew. Chem. Int. Ed. Engl. **15** (1976) 543–544.

(50) E. O. Fischer, E.W. Meineke, F. R. Kreißl, Chem. Ber. **110** (1977) 1140–1147.

(51) E. O. Fischer, G. Besl, Z. Naturforsch. **34b** (1979) 1186–1189.

(52) E. O. Fischer, W. Kleine, W. Schambeck, U. Schubert, Z. Naturforsch. **36b** (1981) 1575–1579.

(53) E. O. Fischer, V. N. Postnov, F. R. Kreißl, J. Organomet. Chem. **231** (1982) C73–C77.

(54) E. O. Fischer, J. Chen, K. Scherzer, J. Organomet. Chem. **253** (1983) 231–241.

(55) E. O. Fischer, J. K. R. Wanner, Chem. Ber. **118** (1985) 2489–2492.

(56) E. O. Fischer, R. L. Clough, P. Stückler, J. Organomet. Chem. **120** (1976) C6–C8.

(57) E. O. Fischer, P. Rustemeyer, D. Neugebauer, Z. Naturforsch. **35b** (1980) 1083–1087.

(58) E. O. Fischer, P. Stückler, H.-J. Beck, F. R. Kreißl, Chem. Ber. **109** (1976) 3089–3098.

(59) E. O. Fischer, J. R. Schneider, D. Neugebauer, Angew. Chem. **96** (1984) 814–815; Angew. Chem. Int. Ed. Engl. **23** (1984) 820.

(60) E. O. Fischer, J. Schneider, J. Organomet. Chem. **295** (1985) C29–C34.

(61) U. Schubert, D. Neugebauer, P. Hofmann, B. E. R. Schilling, H. Fischer, A. Motsch, Chem. Ber. **114** (1981) 3349–3365.

(62) N. M. Kostić, R. F. Fenske, J. Am. Chem. Soc. **103** (1981) 4677–4685.

(63) E. O. Fischer, W. Kleine, F. R. Kreißl, Angew. Chem. **88** (1976) 646–647; Angew. Chem. Int. Ed. Engl. **15** (1976) 616–617.

(64) E. O. Fischer, D. Himmelreich, R. Cai, Chem. Ber. **115** (1982) 84–89.

(65) E. O. Fischer, D. Wittmann, D. Himmelreich, U. Schubert, K. Ackermann, Chem. Ber. **115** (1982) 3141–3151.

(66) E. O. Fischer, D. Wittmann, D. Himmelreich, R. Cai, K. Ackermann, D. Neugebauer, Chem. Ber. **115** (1982) 3152–3166.

(67) U. Schubert, E. O. Fischer, D. Wittmann, Angew. Chem. **92** (1980) 662–663; Angew. Chem. Int. Ed. Engl. **19** (1980) 643–644.

(68) H. Fischer in: *Transition Metal Carbene Complexes:* K. H. Dötz, H. Fischer, P. Hofmann, F. R. Kreißl, U. Schubert, K. Weiss (eds.), Weinheim: Verlag Chemie, 1983; pp. 1–68.

(69) R. Märkl, H. Fischer, J. Organomet. Chem. **267** (1984) 277–284.

(70) A. J. Hartshorn, M. F. Lappert, J. Chem. Soc., Chem. Commun. (1976) 761–762.

(71) A. J. Hartshorn, M. F. Lappert, K. Turner, J. Chem. Soc., Chem. Commun. (1975) 929–930.

(72) J. T. Lin, G. P. Hagan, J. E. Elis, Organometallics **2** (1983) 1145–1150.

(73) E. O. Fischer, D. Himmelreich, R. Cai, H. Fischer, U. Schubert, B. Zimmer-Gasser, Chem. Ber. **114** (1981) 3209–3219.

(74) H. Fischer, E. O. Fischer, D. Himmelreich, R. Cai, U. Schubert, K. Ackermann, Chem. Ber. **114** (1981) 3220–3232.

(75) H. Fischer, E. O. Fischer, R. Cai, Chem. Ber. **115** (1982) 2707–2713.

(76) H. Fischer, E. O. Fischer, R. Cai, D. Himmelreich, Chem. Ber. **116** (1983) 1009–1016.

(77) H. Fischer, A. Motsch, W. Kleine, Angew. Chem. **90** (1978) 914–915; Angew. Chem. Int. Ed. Engl. **17** (1978) 842–843.

(78) H. Fischer, J. Organomet. Chem. **195** (1980) 55–61.

(79) H. Fischer, E. O. Fischer, J. Mol. Catal. **28** (1985) 85–98.

(80) E. O. Fischer, K. Weiß, Chem. Ber. **109** (1976) 1128–1139.

(81) E. O. Fischer, K. Weiß, C. G. Kreiter, Chem. Ber. **107** (1974) 3554–3561.

(82) H. P. Kim, S. Kim, R. A. Jacobson, R. J. Angelici, Organometallics **3** (1984) 1124–1126.

(83) H. P. Kim, R. J. Angelici, Organometallics **5** (1986) 2489–2496.

(84) H. P. Kim, S. Kim, R. A. Jacobson, R. J. Angelici, Organometallics **5** (1986) 2481–2488.

(85) A. Mayr, M. F. Asano, M. A. Kjelsberg, K. S. Lee, D. Van Engen, Organometallics **6** (1987) 432–434.

(86) K. W. Chiu, R. A. Jones, G. Wilkinson, A. M. R. Galas, M. B. Hursthouse, K. M. Abdul Malik, J. Chem. Soc., Dalton Trans. (1981) 1204–1211.

(87) G. R. Clark, K. Marsden, W. R. Roper, L. J. Wright, J. Am. Chem. Soc. **102** (1980) 6570–6571.

(88) W. R. Roper, J. Organomet. Chem. 300 (1986) 167–190.

(89) G. R. Clark, C. M. Cochrane, K. Marsden, W. R. Roper, L. J. Wright, J. Organomet. Chem. **315** (1986) 211–230.

(90) H. Fischer, E. O. Fischer, J. Organomet. Chem. **69** (1974) C1–C3.

(91) D. Himmelreich, E. O. Fischer, Z. Naturforsch. **37b** (1982) 1218.

(92) A. Mayr, G. A. McDermott, A. M. Dorries, Organometallics **4** (1985) 608–610.

(93) A. Mayr, A. M. Dorries, G. A. McDermott, D. Van Engen, Organo-metallics **5** (1986) 1504–1506.

(94) G. A. McDermott, A. M. Dorries, A. Mayr, Organometallics **6** (1987) 925–931.

(95) G. A. McDermott, A. Mayr, J. Am. Chem. Soc. **109** (1987) 580–582.

(96) M. Bottrill, M. Green, J. Am. Chem. Soc. **99** (1977) 5795–5796.

(97) S. R. Allen, R. G. Beevor, M. Green, A. G. Orpen, K. E. Paddick, I. D. Williams, J. Chem. Soc., Dalton Trans. (1987) 591–604.

(98) S. R. Allen, M. Green, A. G. Orpen, I. D. Williams, J. Chem. Soc., Chem. Commun. (1982) 826–828.

(99) R. G. Beevor, M. Green, A. G. Orpen, I. D. Williams, J. Chem. Soc., Dalton Trans. (1987) 1319–1328.

(100) B. D. Dombek, R. J. Angelici, J. Am. Chem. Soc. **97** (1975) 1261–1262.

(101) B. D. Dombek, R. J. Angelici, Inorg. Chem. **15** (1976) 2397–2402.

(102) W. W. Greaves, R. J. Angelici, Inorg. Chem. **20** (1981) 2983–2988.

(103) T. Desmond, F. J. Lalor, G. Ferguson, M. Parvez, J. Chem. Soc., Chem. Commun. (1984) 75–77.

(104) W. W. Greaves, R. J. Angelici, B. J. Helland, R. Klima, R. A. Jacobson, J. Am. Chem. Soc. **101** (1979) 7618–7620.

(105) J. Chatt, A. J. L. Pombeiro, R. L. Richards, G. H. D. Royston, K. W. Muir, R. Walker, J. Chem. Soc., Chem. Commun. (1975) 708–709.

(106) J. Chatt, A. J. L. Pombeiro, R. L. Richards, J. Chem. Soc., Dalton Trans. (1979) 1585–1590.

(107) J. Chatt, A. J. L. Pombeiro, R. L. Richards, J. Chem. Soc., Dalton Trans. (1980) 492–498.

(108) A. J. L. Pombeiro, R. L. Richards, Transition Met. Chem. **5** (1980) 55–59.

(109) J. Chatt, A. J. L. Pombeiro, R. L. Richards, J. Organomet. Chem. **184** (1980) 357–364.

(110) A. J. L. Pombeiro, R. L. Richards, J. R. Dilworth, J. Organomet. Chem. **175** (1979) C17–C18.

(111) A. J. L. Pombeiro, M. F. N. N. Carvalho, P. B. Hitchcock, R. L. Richards, J. Chem. Soc., Dalton Trans. (1981) 1629–1634.

(112) A. J. L. Pombeiro, D. L. Hughes, C. J. Pickett, R. L. Richards, J. Chem. Soc., Chem. Commun. (1986) 246–247.

(113) R. G. Beevor, M. Green, A. G. Orpen, I. D. Williams, J. Chem. Soc., Chem. Commun. (1983) 673–675.

(114) R. G. Beevor, M. J. Freeman, M. Green, C. E. Morton, A. G. Orpen, J. Chem. Soc., Chem. Commun. (1985) 68–70.

(115) K. R. Birdwhistell, T. L. Tonker, J. L. Templeton, J. Am. Chem. Soc. **107** (1985) 4474–4483.

(116) K. R. Birdwhistell, S. J. N. Burgmayer, J. L. Templeton, J. Am. Chem. Soc. **105** (1983) 7789–7790.

(117) A. Höhn, H. Werner, Angew. Chem. **98** (1986) 745–746; Angew. Chem. Int. Ed. Engl. **25** (1986) 737–738.

(118) A. Mayr, K. C. Schaefer, E. Y. Huang, J. Am. Chem. Soc. **106** (1984) 1517–1518.

(119) D. S. Gill, M. Green, J. Chem. Soc., Chem. Commun. (1981) 1037–1038.

(120) N. E. Kolobova, L. L. Ivanov, O. S. Zhvanko, O. M. Khitrova, A. S. Batsanov, Yu. T. Struchkov, J. Organomet. Chem. **262** (1984) 39–47.

(121) N. M. Kostic, R. F. Fenske, Organometallics **1** (1982) 974–982.

(122) N. A. Ustynyuk, V. N. Vinogradova, V. G. Andrianov, Yu. T. Struchkov, J. Organomet. Chem. **268** (1984) 73–78.

(123) P. R. Sharp, S. J. Holmes, R. R. Schrock, M. R. Churchill, H. J. Wasserman, J. Am. Chem. Soc. **103** (1981) 965–966.

(124) K. J. Ahmad, M. H. Chisholm, J. C. Huffman, Organometallics **4** (1985) 1168–1174.

(125) I. Noda, S. Kato, M. Mizuta, N. Yasuoka, N. Kasai, Angew. Chem. **91** (1979) 85–86; Angew. Chem. Int. Ed. Engl. **18** (1979) 83–84.

(126) T. Desmond, F. J. Lalor, G. Ferguson, M. Parvez, J. Chem. Soc., Chem. Commun. (1983) 457–459.

(127) F. G. A. Stone, Angew. Chem. **96** (1984) 85–96; Angew. Chem. Int. Ed. Engl. **23** (1984) 89–100.

(128) F. G. A. Stone, Pure and Appl. Chem. **58** (1986) 529–536.

(129) H. Fischer, B. Bühlmeyer, J. Organomet. Chem. **317** (1986) 187–200.

(130) E. O. Fischer, A. Ruhs, F. R. Kreißl, Chem. Ber. **110** (1977) 805–815.

(131) H. Fischer, A. Ruhs, J. Organomet. Chem. **170** (1979) 181–194.

(132) A. Mayr, M. A. Kjelsberg, K. S. Lee, M. F. Asano, T.-C. Hsieh, Organometallics **6** (1987) 2610–2612.

(133) F. R. Kreißl, J. Organomet. Chem. **99** (1975) 305–308.

(134) E. O. Fischer, A. Ruhs, Chem. Ber. **111** (1978) 2774–2778.

(135) E. O. Fischer, A. Ruhs, P. Friedrich, G. Huttner, Angew. Chem. **89** (1977) 481–482; Angew. Chem. Int. Ed. Engl. **16** (1977) 465–466.

(136) A. Filippou, E. O. Fischer, Z. Naturforsch. **38b** (1983) 587–591.

(137) F. A. Cotton, W. Schwotzer, Inorg. Chem. **22** (1983) 387–390.

(138) H. Fischer, F. Seitz, J. Organomet. Chem. **268** (1984) 247–258.

(139) A. Mayr, A. M. Dorries, G. A. McDermott, S. J. Geib, A. L. Rheingold, J. Am. Chem. Soc. **107** (1985) 7775–7776.

(140) P. G. Byrne, M. E. Garcia, N. Hoa Tran-Huy, J. C. Jeffery, F. G. A. Stone, J. Chem. Soc., Dalton Trans. (1987) 1243–1247.

(141) W. Uedelhofen, K. Eberl, F. R. Kreißl, Chem. Ber. **112** (1979) 3376–3389.

(142) F. R. Kreißl, P. Stückler, E. W. Meineke, Chem. Ber. **110** (1977) 3040–3045.

(143) H. Fischer, A. Motsch, U. Schubert, D. Neugebauer, Angew. Chem. **93** (1981) 483–487; Angew. Chem. Int. Ed. Engl. **20** (1981) 463–464.

(144) H. Fischer, A. Motsch, J. Organomet. Chem. **220** (1981) 301–308.

(145) E. O. Fischer, A. C. Filippou, H. G. Alt, J. Organomet. Chem. **296** (1985) 69–82.

(146) A. C. Filippou, E. O. Fischer, H. G. Alt, J. Organomet. Chem. **310** (1986) 357–366.

(147) E. O. Fischer, D. Wittmann, J. Organomet. Chem. **292** (1985) 245–246.

(148) A. C. Filippou, E. O. Fischer, H. G. Alt, J. Organomet. Chem. **303** (1986) C13–C16.

(149) A. C. Filippou, E. O. Fischer, K. Öfele, H. G. Alt, J. Organomet. Chem. **308** (1986) 11–17.

(150) A. C. Filippou, E. O. Fischer, H. G. Alt, U. Thewalt, J. Organomet. Chem. **326** (1987) 59–81.

(151) E. O. Fischer, A. C. Filippou, H. G. Alt, U. Thewalt, Angew. Chem. **97** (1985) 215–217; Angew. Chem. Int. Ed. Engl. **24** (1985) 203–204.

(152) A. C. Filippou, E. O. Fischer, J. Organomet. Chem. **330** (1987) C1–C4.

(153) H. Fischer, F. Seitz, J. Organomet. Chem. **275** (1984) 83–91.

(154) E. O. Fischer, T. L. Lindner, Z. Naturforsch. **32b** (1977) 713–714.

(155) E. O. Fischer, G. Huttner, T. L. Lindner, A. Frank, F. R. Kreißl, Angew. Chem. **88** (1976) 163–164; Angew. Chem. Int. Ed. Engl. **15** (1976) 157.

(156) E. O. Fischer, P. Friedrich, T. L. Lindner, D. Neugebauer, F. R. Kreißl, W. Uedelhofen, Nguyen Quy Dao, G. Huttner, J. Organomet. Chem. **247** (1983) 239–246.

(157) E. O. Fischer, J. K. R. Wanner, J. Organomet. Chem. **252** (1983) 175–179.

(158) E. O. Fischer, T. L. Lindner, F. R. Kreißl, J. Organomet. Chem. **112** (1976) C27–C30.

(159) E. O. Fischer, T. L. Lindner, G. Huttner, P. Friedrich, F. R. Kreißl, J. O. Besenhard, Chem. Ber. **110** (1977) 3397–3404.

(160) M. Green, J. A. K. Howard, A. P. James, A. N. de M. Jelfs, C. M. Nunn, F. G. A. Stone, J. Chem. Soc., Chem. Commun. (1984) 1623–1625.

(161) M. Green, J. A. K. Howard, A. P. James, C. M. Nunn, F. G. A. Stone, J. Chem. Soc., Dalton Trans. (1986) 187–197.

(162) M. Green, J. A. K. Howard, A. P. James, C. M. Nunn, F. G. A. Stone, J. Chem. Soc., Chem. Commun. (1984) 1113–1114.

(163) M. Green, J. A. K. Howard, A. P. James, C. M. Nunn, F. G. A. Stone, J. Chem. Soc., Dalton Trans. (1987) 61–72.

(164) M. Green, A. G. Orpen, I. D. Williams, J. Chem. Soc., Chem. Commun. (1982) 493–495.

(165) S. J. Holmes, R. R. Schrock, J. Am. Chem. Soc. **103** (1981) 4599–4600.

(166) S. J. Holmes,, D. N. Clark, H.W. Turner, R. R. Schrock, J. Am. Chem. Soc. **104** (1982) 6322–6329.

(167) J. C. Jefferey, J. C.V. Laurie, I. Moore, F. G. A. Stone, J. Organomet. Chem. **258** (1983) C37–C40.

(168) A. P. James, F. G. A. Stone, J. Organomet. Chem. **310** (1986) 47–54.

(169) G. R. Clark, N. R. Edmonds, P. A. Pauptit, W. R. Roper, J. M. Waters, A. H. Wright, J. Organomet. Chem. **244** (1983) C57–C60.

(170) A. E. Bruce, A. S. Gamble, T. L. Tonker, J. L. Templeton, Organometallics **6** (1987) 1350–1352.

(171) A. Mayr, G. A. McDermott, J. Am. Chem. Soc. **108** (1986) 548–549.

(172) E. O. Fischer, A. Ruhs, H. J. Kalder, Z. Naturforsch. **32b** (1977) 473–475.

Solid State Structures of Carbyne Complexes

By Ulrich Schubert

1 Introduction

Up to October 1987 the structures of about 70 carbyne complexes had been determined by X-ray and neutron diffraction (1–62) (Table 1, see the end of this contribution). Due, probably, to the historical development of transition-metal carbyne complex chemistry, group VI metal complexes represent about 80 percent of the known structures. Not surprisingly, therefore, most of the structural information is available for octahedral (or pseudo-octahedral) complexes (e.g. Fig. 1,3–5,7). Tetrahedral coordination geometries have been found in $[(CO)_3ReC]_2SnTTP$ (32) and monomeric complexes of the type $(RO)_3WCR'$ (56, 57). Five-coordinate structures include a unique square-pyramidal complex containing an alkyl, an alkylidene and an alkylidyne ligand at the same tungsten atom (55), a few monomeric trigonal-bipyramidal complexes of iron (Fig. 2) (18) and osmium (28–30), and dimeric complexes $[(RO)_3WCR']_2$, in which a trigonal-bipyramidally coordinated metal center is obtained by the formation of weak alkoxy bridges (Fig. 6) (58, 60). Two structures are known of the four-legged piano-stool type (Fig. 8) (24, 37), and one example each for pentagonal-bipyramidal (36) and capped octahedral geometries (62).

In this article an attempt is made to extract information about bonding from the available structural data of carbyne complexes, therefore structural features will be discussed mainly for octahedral complexes. However, most of the conclusions reached should also be applicable to other types of carbyne complexes. Due to the fact that standard deviations for bond lengths and angles involving light atoms (C, N, O etc.) are rather high in most cases (because of the presence of heavy atoms like Br, I, W etc.), only trends for the correlation between the chemical environment of a carbyne ligand and its structural parameters can be extracted.

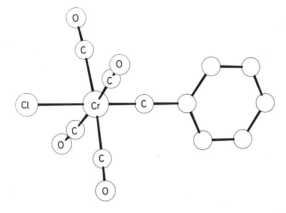

Figure 1. *trans*-Cl(CO)$_4$CrCPh (2). The phenylcarbyne ligand, Cr and Cl are located on a crystallographic mirror plane which bisects the Cr(CO)$_4$ moiety.

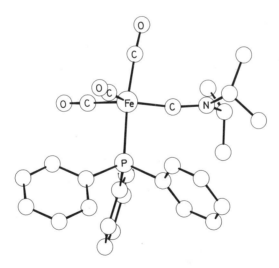

Figure 2. Trigonal-bipyramidal [(PPh$_3$)(CO)$_3$FeCNPr$_2^i$] BCl$_4$ (18). The anion is omitted in the drawing. The phosphine ligand occupies an axial and the carbyne ligand an equatorial position. The plane of the amino substituent roughly coincides with the equatorial plane.

2 The Carbyne Ligand

2.1 M-C Bond Lengths

M-C(carbyne) bond lengths are typically between 165 and 175 pm for first-row transition metals (Cr, Mn, Fe), and between 175 and 190 pm for second- and third-row transition metals (Mo, W, Os, Ta). There is no doubt that these are the shortest known metal-carbon distances. Much effort, however, has been spent in comparing the experimentally found distances with "theoretical" metal-carbon triple bond lengths. Assigning a single "covalent radius" to a metal atom irrespective of its particular ligand sphere can only lead to rough estimates. To test the hypothesis that metal-C(carbyne) distances are equivalent to the sum of the triple-bond radii of carbon and the corresponding metal atom, a comparison of carbyne complexes, $L_nM \equiv CR$, with both acetylenes, $RC \equiv CR$, and dinuclear complexes, $L_nM \equiv ML_n$, in which the same L_nM fragments are connected by a $M \equiv M$ triple bond, would be appropriate. Unfortunately, this comparison can only be made for $Cp(CO)_2Cr \equiv C\text{-}CPh = CPh\text{-}C \equiv Cr(CO)_2Cp$ (17) and for $(RO)_3W \equiv CR'$ complexes (56), due to the lack of other suitable dinuclear complexes containing $M \equiv M$ bonds. Typical $W \equiv W$ triple bond lengths in W_2X_6 are in the range of 225–232 pm (63). Taking half this value and adding the conventional triple-bond radius of carbon (60.3 pm) gives approximately the W-C(carbyne) distance found in $(RO)_3W \equiv CR'$ and its dimers (175–177 pm) (56– 58, 60). Similarly, the $Cr \equiv C$ distance in the $Cp(CO)_2Cr$-carbyne complex (170.7(2) pm) (17) corresponds to half the $Cr \equiv Cr$ bond length in $Cp_2Cr_2(CO)_4$ (220.0(3) and 223.0(3) pm, two independent molecules (64)) plus half the $C \equiv C$ bond length in acetylenes ($C \equiv C$ about 121 pm).

2.2 Conjugation of the $M \equiv C$ Bond with π-Donating Substituents

For a description of structural phenomena in Fischer-type *carbene* complexes the carbene carbon can be considered as an sp^2-hybridized carbenium center, whose positive charge is diminished by π-bonding with its substituents (the metal complex moiety, L_nM, and the two "organic" substituents). The degree of π-bonding of any of these groups, which can be judged from its bond distance to the carbene carbon, depends not only on its own π-donating ability, but also on the π-donor properties of the other groups. Thus, the three groups attached to the carbene carbon compete with each other for π-bonding to the formally empty p-orbital on the carbene carbon (65). If this notion is applied to *carbyne* complexes, π-bonding between the carbyne carbon atom and one of its substituents (L_nM or X) will change, if the π-donating properties of the other substituent are altered.

$$L_nM \equiv C\text{-}X \longleftrightarrow L_n\overset{-}{M} = C = \overset{+}{X}$$

(A)   (B)

A strong contribution of the heteroallenic form (B) to the bonding situation should result in a lengthening of the MC bond and a shortening of the CX bond. Amino groups ($X = NR_2$) are strongly π-donating, and there are abundant data on aminocarbyne complexes. A rough inspection of Table 1 shows that MC(carbyne) distances in aminocarbyne complexes, L_nMCNR_2, generally are some pm longer than M-C(carbyne) distances in complexes having the same metal fragment (L_nM) but a carbyne ligand without a π-donating substituent. In all aminocarbyne complexes (Fig. 2-5) the amino group is planar, thus meeting the requirement for participation in a π-bond with the carbyne carbon, and the C(carbyne)-N distance is shorter than a single bond. A systematic study on bonding in aminocarbyne complexes has been performed for a number of *trans*-$Y(CO)_4CrCNEt_2$ complexes with different ligands Y (12). The longest Cr-C(carbyne) (179.7(9) pm) and the shortest C(carbyne)-N distances (125.6(12) pm) are experimentally found within this series in $[(CO)_5CrCNEt_2]^+$ (i.e. Y = CO). As the π-accepting CO ligand trans to C(carbyne) is successively replaced by PPh_3 (12), $SnPh_3^-$ (15), $SePh^-$ (16) or the π-donating Br^- (13), the metal fragment $Y(CO)_4Cr$ becomes a better π-donor towards the carbyne ligand and the influence of mesomeric form (B) is diminished. The Cr-C(carbyne) distance therefore shortens and the C(carbyne)-N distance lengthens in that order (in *trans*-$Br(CO)_4CrCNEt_2$ (13) Cr-C is 172(1) pm and C-N 129.4(12) pm). In other words, the $Y(CO)_4Cr$ fragment can compete more successfully with the amino group in π-bonding to the carbyne carbon, if $Y = Br^-$ as opposed to $Y = SePh^-$, $SnPh_3^-$, PPh_3 or CO (in that order).

In any octahedral carbyne complex the M-C(carbyne)bond length is influenced by the nature of the trans-ligand, although its effect is often hidden by relatively large standard deviations. Generally, the shortest $M \equiv C$ bonds are found when strongly π-donating ligands, like cyclopentadienyl ligands, are trans to the carbyne ligand. Cis ligands also affect the $M \equiv C$ bond. Their electronic influence, however, is overlaid by steric effects. Since both effects may be opposed, the impact of cis-ligands on the bonding parameters of the carbyne ligands is often difficult to analyse from structural data. For instance, in a geometrically undistorted *mer*-$Br(PPh_3)(CO)_3CrCNEt_2$ (PPh_3 cis to $CNEt_2$, Fig. 3) the Cr-C(carbyne) distance should be shorter and the C(carbyne)-N distance longer than in *trans*-$Br(CO)_4CrCNEt_2$, according to MO calculations (12). The contrary has been found experimentally. Weakening of the Cr-C(carbyne) bond is caused by severe geometrical distortions within the coordination plane at chromium containing the carbyne ligand and both the bromine and phosphorus atom (12).

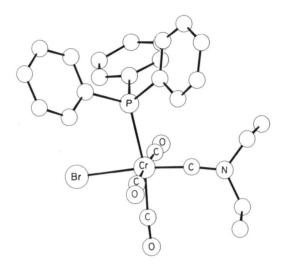

Figure 3. *mer*-Br(PPh$_3$)(CO)$_3$CrCNEt$_2$ (12). C(carbyne)-Cr-P 99.8(4)°, Br-Cr-C(CO) 82.5(4), C(carbyne)-Cr-Br 172.1(4). The dihedral angle between the plane of the amino substituent and the C(carbyne), Cr, P plane is 12°.

Apart from aminocarbyne complexes, the number of carbyne complexes L$_n$M ≡ C-X, in which X is a (potentially) π-donating substituent, whose structures have been investigated, is small. In Cp(PPh$_3$)(CO)W ≡ C-SPh (50) a major contribution by a heteroallenic resonance form was excluded, because the C(carbyne)-S distance (171.6(10) pm) is close to the value expected for a C(sp)-S single bond and because the W ≡ C distance is the same as in Cp(CO)$_2$W ≡ CTol (52) and Cp(CO)$_2$W ≡ CSiPh$_3$ (51). A C(carbyne)-S single bond (171.2(4) pm) is also observed in (HTPB)(CO)$_2$Mo ≡ C-S-*p*-C$_6$H$_4$NO$_2$ (HTPB = hydrotris-(3,5-dimethyl-1-pyrazolyl)borate) (27). Probably due to the strongly electron-donating Cp(PPh$_3$)(CO)W or (HTPB)(CO)$_2$Mo fragments, the S-aryl groups cannot (or need not) compete for π-bonding to the carbyne carbon (similar phenomena have been observed in Cp(CO)$_2$Mn-carbene complexes (66)). In (TPB)(CO)$_2$Mo ≡ CCl (TPB = tetrakis(1-pyrazolyl)borate) (26) the Mo-C(carbyne) distance is significantly longer (189.4(10) pm) than in the related complex (HTPB)(CO)$_2$Mo ≡ CSR (180.1(4) pm). The rather short C(carbyne)-

Cl bond length (154.7(11) pm; in chloroacetylenes C(sp)-Cl distances of about 163 pm have been found (67)) indicates that Cl is a stronger π-donor than SR and that $L_n\overset{-}{M}o = C = \overset{+}{C}l$ contributes strongly to bonding.

The question, whether $M \equiv C$ bonds can be conjugated with organic multiple bonds, is hard to decide from structural parameters, although MO calculations indicate that there should be some conjugation even with aryl substituents (68).

The changes in C-C bond orders induced by such an interaction are too small to result in significant bond length alterations. Remarkably, a high precision structure determination and electronic deformation density determination of $Cl(CO)_4Cr \equiv CMe$ revealed, that even a methyl group can act as an electron donor to the $M \equiv C$ bond by hyperconjugation (2). This interpretation has been confirmed by MO calculations (69). Hyperconjugation should be stronger in silylcarbyne complexes (68).

2.3 Conformational Effects

The CH ligand is characterized by two degenerate p-orbitals which are orthogonal to each other. In CR ligands degeneration of the two acceptor orbitals is lifted, if the substituent R has no rotational symmetry. For R = aryl or R = alkyl the difference in energy between the two orbitals remains low, and two nearly equivalent π-interactions with suitable metal complex moieties can develop. In aminocarbyne (and related) ligands, splitting between the two acceptor orbitals is large, due to interaction with the filled nitrogen p-orbital. The aminocarbyne ligand therefore electronically resembles a vinylidene ligand (12). Because of this "single-faced" character, a preferred orientation of the aminocarbyne ligand must result, if the donor orbitals of the L_nM fragment have no rotational symmetry. Conformational preferences are typical in the structural chemistry of carbene complexes (65), however, there is an important difference: in *carbene* ligands the acceptor orbital is *perpendicular* to the carbene plane, while in *aminocarbyne* ligands the dominating acceptor orbital is *within* the plane of the planar $C-NR_2$ ligand.

In octahedral carbyne complexes of the type *trans*-$X(CO)_4M \equiv C-NR_2$, in which the trans-ligand X has rotational symmetry (e.g. X = halide, CO), there is no electronic preference for a certain orientation of the NR_2 plane. The crystallographically found dihedral angles C(CO)-M-N-R are caused by steric repulsion between the amino substituent and the equatorial ligands and by packing effects. An interesting situation arises, if the trans-ligand X has no rotational symmetry, as in *trans*-*p*-$FC_6H_4Se(CO)_4Cr \equiv NEt_2$ (16). Due to the "single-faced" donor character of the PhSe ligand, the donor orbitals of the $PhSe(CO)_4Cr$ moiety are no longer degenerate. The electronically preferred interaction with the aminocarbyne ligand places the amino substituent in a plane perpendicular to the Cr, Se, C(phenyl) plane (Fig. 4) and approximately coplanar with two of the cis-CO ligands.

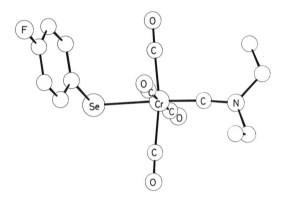

Figure 4. *trans-p*-FC₆H₄-Se(CO)₄CrCNEt₂ (16). The plane of the amino substituent is perpendicular to the Cr, Se, C(phenyl) planes and approximately coplanar with two CO ligands.

More frequently, symmetry of the trans-X(CO)₄M moiety is lowered by substitution of one CO ligand. MO calculations predict (12), that the entering ligand (L) should be located within the plane of the aminocarbyne ligand, if L is a better electron donor than CO. This orientation of the amino plane (A) is electronically preferred by a few kJ. However, the alternate orientation (B) is sterically favoured. In reality, a compromise is reached by twisting the amino plane out of conformation (A) by some degrees: 12° in *mer*-Br(PPh₃)(CO)₃-CrCNEt₂ (12), and 8°–21° in dimeric complexes of the type (*C*) (M = Mo, W) (23).

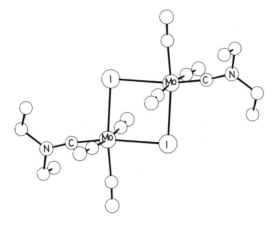

Figure 5. [(μ-I)(CO)₃MoCNEt₂]₂ (23). The dihedral angle between the plane of the amino substituent and the Mo₂I₂ plane is 21°. The Mo₂I₂ unit is planar by space group symmetry.

No carbyne complex of the type $(\pi\text{-arene})(CO)_2M \equiv CNR_2$ has been structurally characterized; depending on the $(\pi\text{-arene})M$ unit, the amino plane should either be parallel or perpendicular to the mirror plane of the $(\pi\text{-arene})(CO)_2M$ moiety. If one of the CO ligands is substituted by a better donor ligand (L), the dominating acceptor orbital of the carbyne ligand should be coplanar with the L, M, C(carbyne) plane (70). This orientation is indeed found in $Cp(PPh_3)(CO)W \equiv CSPh$ (50).

There are two structurally characterized monomeric complexes with a trigonal-bipyramidal geometry which contain a "single-faced" carbyne ligand. In each of them, $(PPh_3)(CO)_3FeCNPr_2$ (18) and $[(PPh_3)_2(CO)_2OsCTeMe]^+$ (30), the phosphine ligand is in the apical and the carbyne ligand is in the equatorial position. Both the NR_2 and the TeMe substituent are located within the equatorial plane (Fig. 2) in agreement with theoretical predictions (71). Complexes of the type $(OR)_3WCR$ form weakly associated dimers, in which the metal atom has a trigonal-bipyramidal geometry (58, 60). The carbyne ligand and the weaker of the two W-O(bridge)bonds occupy the apical positions (Fig. 6). In dimeric $(OBu^t)_3WCNMe_2$ (60) the amino plane is perpendicular to the W_2O_2 plane.

As the splitting of the two π-acceptor orbitals of the carbyne ligand decreases, the energetic advantage of a conformation which maximises π-overlap becomes less important than steric considerations. Although the two π-accepting orbitals of a phenylcarbyne are not equivalent, their difference in energy is low (68, 72).

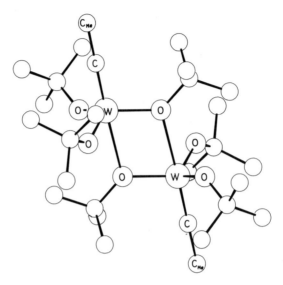

Figure 6. Dimeric (OBu^t)$_3$WCMe (58). In the planar W$_2$O$_2$ unit the extremely long W-O distances trans to the carbyne ligand (248.4(4)pm; for comparison: W-O(equatorial) 193.4 (4)pm) indicate that alkoxy bridge formation is weak. Hydrogen atoms have been omitted for clarity.

Therefore, in the structures of aryl- (and, similarly, vinyl-) substituted carbyne complexes no preferred orientation of the aryl (vinyl) plane is observed. In many cases the phenyl plane is located between the cis-ligands at the metal, as it is ideally found in X(CO)$_4$CrCPh (X = Cl, Br) (2) (Fig. 1).

3 The Metal Complex Moiety

Carbyne ligands are among the strongest π-accepting ligands and therefore strongly influence both the geometry of the complexes and the bonding of other ligands coordinated to the same metal. If several *isomers* of a carbyne complex are possible, the carbyne ligand will strive for a coordination site in which it can accept a maximum amount of electron density. In octahedral complexes this is usually trans to the ligand which is the strongest π-donor, typically a halide, or the weakest π-acceptor. In trigonal-bipyramidal structures of d^8 complexes a strong π-acceptor, like the carbyne ligand, will tend to occupy an equatorial site (71) (see Fig. 2).

Bonding of the *trans-ligand* is strongly affected by the π-accepting properties of a carbyne ligand. Metal-ligand bond distances increase appreciably if the particular ligand is trans to a carbyne ligand. For instance, in $[(CO)_5CrCNEt_2]^+$ Cr-C(CO) trans to $CNEt_2$ is one of the longest Cr-C(CO)distances (198(1) pm) ever observed (12).

The magnitude of this effect is particularly well demonstrated in complexes having the same ligand both cis and trans to the carbyne ligand. Complexes of the type

$$\begin{array}{c} X \\ | \, / \\ X-M\equiv C-R \\ / \, | \\ C \\ O \end{array} \quad \text{or} \quad \begin{array}{c} X \ \ X \\ | \, / \\ X-M\equiv C-R \\ / \, | \\ {}_{O}{}^{C} \ \ {}^{C}{}_{O} \end{array}$$

allow a comparison of CO and CR ligands with respect to their trans-influence on the M-X bonds. In $[(CO)_5CrCNEt_2]^+$ (X = CO) the difference in bond distances between the CO ligands cis and trans to CR [Δ(M-CO)] is 11 pm! Similary, in complexes of the type (C) [Δ(M-X)] is 6 pm (M = Mo, X = I), 4 pm (M = Mo, X = NCO), 1 pm (M = W, X = N_3; not significant) and 7 pm (M = W, X = SPh), respectively (23). In cyclopentadienyl or the related polypyrazolylborate complexes, in which the X_3 unit is occupied by the polydentate ligand, the M-C or M-N distance trans to CR is longer than the other ones, i.e. the cyclopentadienyl or polypyrazolylborate ligand is somewhat tilted in respect to the metal fragment. In $(HTPB)(CO)_2Mo \equiv CSC_6H_4NO_2$ (27) and $(TPB)(CO)_2$-$Mo \equiv CCl$ (26) [Δ(Mo-N)] is 4–8 pm, while in Cp-substituted carbyne complexes, $Cp(CO)_2M \equiv CR$, the difference between the longest M-C(Cp) distance (trans to CR) and the shortest one (trans to CO) is typically in the range of 3 pm for Cr and Mn and 6 pm for Mo and W. The longest Ta-C distance (251.8(9) pm) in $(C_5Me_5)(PMe_3)_2ClTaPh$ (Fig. 7) is again found for the carbon atom which is approximately trans to CPh, while the shortest one (239.7(9) pm), not surprisingly, is trans to Cl (37). In $(py)_2I_2(CHBu^t)Re \equiv CBu^t$ (35) an internal comparison between the trans-influence of a carbene and a carbyne ligand is possible, since both are trans to a pyridine ligand (and cis to each other), thus occupying equivalent sites in an octahedral complex. The Re-N distance trans to CBu^t is 4.6 pm longer than Re-N trans to $CHBu^t$, showing that the carbyne ligand is a stronger π-acceptor than the carbene ligand.

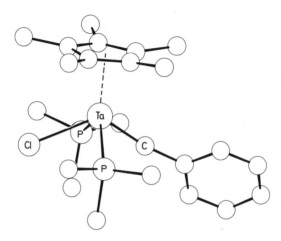

Figure 7. $(C_5Me_5)(PMe_3)_2ClTaCPh$ (37). C(carbyne)-Ta-P 80.3(2), 79.0(2), Cl-Ta-P 76.66(7), 75.76(7), C(carbyne)-Ta-Cl 125.5(2)°.

The influence of the π-accepting carbyne ligand on *cis-ligands* seems to be less pronounced. An analysis of such effects is more difficult since electronic and steric effects overlap, as has already been discussed for *mer*-Br(PPh₃)(CO)₃-CrCNEt₂ (see above). In *trans*-X(CO)₄M $\equiv$ CR the cis-CO ligands are generally bent away from CR by up to 10° even if X is a bulky ligand (e.g. in Ph₃Sn-(CO)₄CrCNEt₂ (15) one of the C(CO)-Cr-C(carbyne) angles is 100.9(7)°), although individual bond angles may vary considerably and may even be smaller than 90°. The same is true for complexes of the type *trans*-XL₄M $\equiv$ CR, L being a phosphine or phosphite ligand.

In *trans*-*p*-FC₆H₄Se(CO)₄Cr $\equiv$ CNEt₂ (16) (Fig. 4) the carbonyl ligands are pairwise different. The two CO ligands which are located approximately within the plane of the aminocarbyne ligand have longer Cr-C distances (mean 194 pm) and smaller Se-Cr-C(CO)angles (mean 85.0°) than the others (190 pm, 89.8°).

This is probably a consequence of the fact that the two Cr-C π-interactions are not equivalent (see chapter 2.3). Lowering of the symmetry of the metal fragment by interaction with a "single-faced" carbyne ligand should be quite general, but little attention has been paid to it and the corresponding structural parameters are not always reported. The phenomenon of symmetry reduction has also been observed in a high-precision structure determination of *trans*-Cl(CO)$_4$Cr ≡ CMe (2) to a surprisingly great extent. In the crystalline state one of the hydrogen atoms of the methyl group is found to lie in the plan of two carbonyl groups, the Cr-C distances of which are significantly shorter (192.6 and 192.9 pm) than the distances of the CO ligands perpendicular to them (198.1 and 198.7 pm). A deformation density map of this compound shows the same reduction of symmetry, the higher electron density being in the longer metal-carbon bonds (2). Since a rather high electron density was found in the C(carbyne)-C(methyl)bond, one possible explanation for this feature could be a hyperconjugative interaction of the methyl group with the metal complex fragment. However, the particular packing of the molecules could also be responsible for the observed distortions, since rather short distances between the hydrogen atoms of the methyl group and the chlorine ligands of neighbouring molecules are observed (266–277 pm).

Table 1. Compilation of structurally characterized carbyne complexes with important bond lengths (in pm) (n.r. = not reported; only a significant number of digits is given).

Complex	M-C(carbyne)	M-X(trans to C(carbyne))	Ref.
trans-I(CO)$_4$CrCMe	169(1)	I: 279.2(2)	(1)
trans-Cl(CO)$_4$CrCMe	171.0	Cl: 244.2	(2)
trans-Br(CO)$_4$CrC(*p*-C$_6$H$_4$CF$_3$)	168(2)	Br: 256.3(4)	(3)
trans-Cl(CO)$_4$CrCPh	168(1)	Cl: 240.4(4)	(4)
	172.8(4)	241.4(4)	(5)
trans-Br(CO)$_4$CrCPh	168(3)	Br: 256.7(3)	(4)
trans-I(CO)$_4$CrC(C$_5$H$_7$)	165(2)	I: 278.1(3)	(6)
trans-Br(CO)$_4$CrC(C$_5$H$_4$FeCp)	171(2)	Br: 257.7(3)	(7)
(−)*trans*-Br(CO)$_4$CrC(menthyl)	167	Br: 256	(8)
trans-(CO)$_5$CrBr(CO)$_4$CrC(C$_3$H$_4$)	171(1)	Br: 257.6(2)	(9)
mer-Br(PMe$_3$)(CO)$_3$CrCMe	168(3)	Br: 260.3(5)	(1)

continue

Table 1 Continued.

Complex	M-C(carbyne)	M-X(trans to C(carbyne))	Ref.
trans,cis-Br(CNBu^t)$_2$(CO)$_2$CrCPh	176(3)	Br: 274.8(5)	(10)
trans,trans-Br[P(OPh)$_3$]$_2$(CO)$_2$CrCPh			
	168(1)	Br: 254.4(2)	(10)
[*trans*-(PMe$_3$)(CO)$_4$CrCMe]BCl$_4$	167(1)	P: 247.4(4)	(11)
[(CO)$_5$CrCNEt$_2$]BF$_4$	179.7(9)	C: 198(1)	(12)
[*trans*-(PPh$_3$)(CO)$_4$CrCNEt$_2$]BF$_4$	176(1)	P: 246.4(3)	(12)
trans-Br(CO)$_4$CrCNEt$_2$	172(1)	Br: 256.4(2)	(13)
trans-Cl(CO)$_4$CrCNPr$_2^i$	174.7(5)	Cl: 241.2(1)	(14)
trans-(SnPh$_3$)(CO)$_4$CrCNEt$_2$	174(1)	Sn: 271.9(2)	(15)
trans-(Se-*p*-C$_6$H$_4$F)(CO)$_4$CrCNEt$_2$	175(1)	Se: 256.2(2)	(16)
mer-Br(PPh$_3$)(CO)$_3$CrCNEt$_2$	175(1)	Br: 257.2(2)	(12)
[Cp(CO)$_2$CrC-C(Ph)=]$_2$	170.7(2)	C(Cp): 223.1(2) 222.8(2)	(17)
[*cis*-(PPh$_3$)(CO)$_3$FeCNPr$_2^i$]BCl$_4$	173.4(6)	–	(18)
[Cp(CO)$_2$MnC-CH = CPh$_2$]BF$_4$	166.5(5)	C(Cp): 216.2(6) 216.3(6)	(19)
trans-(CO)$_5$Re(CO)$_4$MoCPh	184(3)	Re: 311.1(2)	(20)
trans-Br(dppe)$_2$MoCSiMe$_3$	182(1)	Br: 273.1(2)	(21)
trans-Cl[P(OMe)$_3$]$_4$MoCPh	179.3(8)	Cl: 258.3(3)	(22)
[*mer*-(μ-I)(CO)$_3$MoCNEt$_2$]$_2$	179(2)	I: 294.3(2)	(23)
[*mer*-(μ-NCO)(CO)$_3$MoCNEt$_2$]$_2$	182(2)	N: 229(1)	(23)
[Cp[P(OMe)$_3$]$_2$(H)MoCCH$_2$Bu^t]BF$_4$	179.8(2)	n. r.	(24)
Cp[P(OMe)$_3$]$_2$MoCCH$_2$Bu^t	179.6(2) 179.9(2)	C(Cp): 242.7(3) 242.5(3) 241.8(3)	(25)
[B(C$_3$H$_3$N$_2$)$_4$](CO)$_2$MoCCl	189.4(9) 189.5(10) 189.4(10)	N: 223.5(7) 225.0(7) 226.3(7)	(26)
[HB(C$_3$HMe$_2$N$_2$)$_3$](CO)$_2$MoCS(*p*-C$_6$H$_4$NO$_2$)			
	180.1(4)	N: 229.0(4)	(27)

continue

Table 1 Continued.

Complex	M-C(carbyne)	M-X(trans to C(carbyne))	Ref.
(PPh₃)₂(Cl)(CO)OsCTol	178(2)	–	(28, 29)
[(PPh₃)₂(CO)₂OsCTeMe]⁺	184(2)	–	(30)
trans,trans-(PPh₃)₂Cl₂(NCS)OsC(*p*-C₆H₄NMe₂)			
	175(1)	N: n. r.	(31)
[*trans,trans*-(PPh₃)₂Cl₂(CNTol)OsC(*p*-C₆H₄NMe₂)]ClO₄			
	178(1)	C: 214(1)	(31)
[(CO)₃ReC]₂SnTTP	175	–	(32)
[*trans*-Cl(dppe)₂ReCNH₂]BF₄	180.2(4)	Cl: 248.5(1)	(33)
[*trans*-Cl(dppe)₂ReCNHMe]BF₄	180(3)	Cl: 248.4(6)	(34)
(py)₂I₂(CHBuᵗ)ReCBuᵗ	174.2(9)	N: 241.5(6)	(35)
Me₃AlCl(dmpe)₂(H)TaCBuᵗ	185.0(5)	Cl: 275.8(2)	(36)
(C₅Me₅)(PMe₃)₂ClTaCPh	184.9(8)	C: 251.8(9)	(37)
trans-X(CO)₄WCMe (X = Br,I)	n. r.	n. r.	(38)
trans-Cl(CO)₄WCMe	202(4)	Cl: 248(1)	(11, 38)
trans-I(CO)₄WCPh	190(5)	I: 284.5(5)	(1)
trans-Cl(CO)₄WC[C₆H₅Cr(CO)₃]	184(3)	Cl: 248.5(8)	(39)
trans-(CO)₄Co(CO)₄WCPh	182(1)	Co: 294.0(2)	(40)
[*trans*-Br(CO)₄WC]₂(*p*-C₆H₄)	189(3)	Br: 262.7(5)	(41)
[*mer*-(μ-N₃)(CO)₃WCNEt₂]₂	175(4)	N: 223(3)	(23)
[*mer*-(μ-SPh)(CO)₃WCNEt₂]₂	184(3)	S: 259.8(7)	(23)
	180(2)	262.1(6)	
[*fac*-(μ-Cl)(Cl)(PMe₃)₂WCPMe₃]₂(AlCl₄)₂			
	183(3)	Cl: 254.5(8)	(42)
trans,cis-Br(CO)₂(py)₂WCPh	184(2)	Br: 269.6(2)	(43)
trans,cis-Cl(CO)(py)₂(C₄H₂O₃)WCPh			
	180.1(6)	Cl: 253.8(1)	(44)
trans-Cl(PMe₃)₄WCH	184	Cl: 241.9(2)	(45)
	184	244.2(5)	
trans-Me(PMe₃)₄WCMe	189(3)	C: 245(3)	(46)
[(CO)₅WCNEt₂]SbCl₆	190(3)	C: 216(3)	(47)
	180(3)	210(3)	

continue

Table 1 Continued.

Complex	M-C(carbyne)	M-X(trans to C(carbyne))	Ref.
$(CO)_5Cr$-Ph_2P-$(CO)_2(bipy)WCNEt_2$	187.7(8)	N: 225.0(6)	(48)
trans-$Br(CO)_4WCNCPh_2$	187.8(5)	Br: 264.4(1)	(49)
$Cp(PPh_3)(CO)WCSPh$	180.7(10)	C(Cp): 238.5(10)	(50)
$Cp(CO)_2WCSiPh_3$	181(2)	n. r.	(51)
$Cp(CO)_2WCTol$	182(2)	C(Cp): 238[a]	(52)
$[(C_5H_4Bu^t)(I)WCBu^t]_2N_2H_2$	176.9(8)	C(Cp): 247.3(9)	(53)
$[B(C_3H_3N_2)_4](CO)_2WCTol$	182.1(7)	N: 228.4(6)	(54)
$(dmpe)(CH_2Bu^t)(CHBu^t)WCBu^t$	178.5(8)	–	(55)
$(OBu^t)_3WCPh$	175.8(5)	–	(56)
$(OBu^t)_3WCRu(CO)_2Cp$	175(2)	–	(57)
$[(OBu^t)_2(\mu\text{-}OBu^t)WCMe]_2$	175.9(6)	O: 248.4(4)	(58)
$[(OPr^i)_2(\mu\text{-}OPr^i)(NHMe_2)WCEt]_2$	177(1)	O: 236.0(7)	(59)
$[(OBu^t)_2(\mu\text{-}OBu^t)WCNMe_2]_2$	177(2)	O: 242(1)	(60)
	175(2)	243(1)	
$Cl_2(PEt_3)_2(PHPh)WCBu^t$	180.8(6)	Cl: 257.8(2)	(61)
$Cl_3(PMe_3)_3WCBu^t$	179.3(6)	–	(62)

[a] derived from a calculated position

4 References

(1) G. Huttner, H. Lorenz, W. Gartzke, *Angew. Chem.* **86** (1974) 667–669; *Angew. Chem. Int. Ed. Engl.* **13** (1974) 609–610.

(2) C. Krüger, R. Goddard, K. H. Claus, *Z. Naturforsch.* **38b** (1983) 1431–1440.

(3) E. O. Fischer, A. Schwanzer, H. Fischer, D. Neugebauer, G. Huttner, *Chem. Ber.* **110** (1977) 53–66.

(4) A. Frank, E. O. Fischer, G. Huttner, *J. Organomet. Chem.* **161** (1978) C27–C30.

(5) N. Q. Dao, D. Neugebauer, H. Fevrier, E. O. Fischer, P. J. Becker, J. Pennetier, *Nov. J. Chim.* **6** (1982) 359–364.

(6) E. O. Fischer, W. R. Wagner, F. R. Kreißl, D. Neugebauer, *Chem. Ber.* **112** (1979) 1320–1328.

(7) E. O. Fischer, M. Schluge, J. O. Besenhard, P. Friedrich, G. Huttner, F. R. Kreißl, *Chem. Ber.* **111** (1978) 3530–3541.

(8) S. Fontana, O. Omara, E. O. Fischer, U. Schubert, F. R. Kreißl, *J. Organomet. Chem.* **149** (1978) C57–C62.

(9) E. O. Fischer, N. H. Tran-Huy, D. Neugebauer, *J. Organomet. Chem.* **229** (1982) 169–177.

(10) A. Frank, U. Schubert, G. Huttner, *Chem. Ber.* **110** (1977) 3020–3025.

(11) G. Huttner, A. Frank, E. O. Fischer, *Israel J. Chem.* **15** (1976/77) 133–142.

(12) U. Schubert, D. Neugebauer, P. Hofmann, B. E. R. Schilling, H. Fischer, A. Motsch, *Chem. Ber.* **114** (1981) 3349–3365.

(13) E. O. Fischer, W. Kleine, G. Kreis, F. R. Kreißl, *Chem. Ber.* **111** (1978) 3542–3551; E. O. Fischer, G. Huttner, W. Kleine, A. Frank, *Angew. Chem.* **87** (1975) 781; *Angew. Chem. Int. Ed. Engl.* **14** (1975) 760.

(14) H. Fischer, A. Motsch, R. Märkl, K. Ackermann, *Organometallics* **4** (1985) 726–735.

(15) U. Schubert, *Cryst. Struct. Comm.* **9** (1980) 383–388; E. O. Fischer, H. Fischer, U. Schubert, R. B. A. Pardy, *Angew. Chem.* **91** (1979) 929–930; *Angew. Chem. Int. Ed. Engl.* **18** (1979) 871.

(16) H. Fischer, E. O. Fischer, D. Himmelreich, R. Cai, U. Schubert, K. Ackermann, *Chem. Ber.* **114** (1981) 3220–3232.

(17) N. A. Ustynynk, V. N. Vinogradova, V. G. Andrianov, Yu. T. Struchkov, *J. Organomet. Chem.* **268** (1984) 73–78.

(18) E. O. Fischer, J. Schneider, D. Neugebauer, *Angew. Chem.* **96** (1984) 814–815; *Angew. Chem. Int. Ed. Engl.* **23** (1984) 820–821.

(19) N. E. Kolobova, L. L. Ivanov, O. S. Zhvanko, O. M. Khitrova, A. S. Batsanov, Yu. T. Struchkov, *J. Organomet. Chem.* **262** (1984) 39–47.

(20) E. O. Fischer, G. Huttner, T. L. Lindner, A. Frank, F. R. Kreißl, *Angew. Chem.* **88** (1976) 163–164; *Angew. Chem. Int. Ed. Engl.* **15** (1976) 157.

(21) K. J. Ahmed, M. H. Chisholm, J. C. Huffman, *Organometallics,* **4** (1985) 1168–1174.

(22) A. Mayr, A. M. Dorries, G. A. McDermott, D. van Engen, *Organometallics* **5** (1986) 1504–1506.

(23) E. O. Fischer, D. Wittmann, D. Himmelreich, R. Cai, K. Ackermann, D. Neugebauer, *Chem. Ber.* **115** (1982) 3152–3166.

(24) M. Green, A. G. Orpen, I. D. Williams, *J. Chem. Soc. Chem. Comm.* (1982) 493–495.

(25) S. R. Allen, R. G. Beevor, M. Green, A. G. Orpen, K. E. Paddick, I. D. Williams, *J. Chem. Soc. Dalton* (1987) 591–604.

(26) T. Desmond, F. J. Lalor, G. Ferguson, M. Parvez, *J. Chem. Soc. Chem. Comm.* (1983) 457–459.

(27) T. Desmond, F. J. Lalor, G. Ferguson, M. Parvez, *J. Chem. Soc. Chem. Comm.* (1984) 75–77.

(28) G. R. Clark, K. Marsden, W. R. Roper, L. J. Wright, *J. Am. Chem. Soc.* **102** (1980) 6570–6571.

(29) G. R. Clark, C. M. Cochrane, K. Marsden, W. R. Roper, L. J. Wright, *J. Organomet. Chem.* **315** (1986) 211–230.

(30) W. R. Roper, *J. Organomet. Chem.* **300** (1986) 167–190.

(31) G. R. Clark, N. R. Edmonds, R. A. Pauptit, W. R. Roper, J. M. Waters, A. H. Wright, *J. Organomet. Chem.* **244** (1983) C57–C60.

(32) I. Noda, S. Kato, M. Mizuta, N. Yasuoka, N. Kasai, *Angew. Chem.* **91** (1979) 85–86; *Angew. Chem. Int. Ed. Engl.* **18** (1979) 83.

(33) A. J. L. Pombeiro, D. L. Hughes, C. J. Pickett, R. L. Richards, *J. Chem. Soc. Chem. Comm.* (1986) 246–247.

(34) A. J. L. Pombeiro, M. F. N. N. Carvalho, P. B. Hitchcock, R. L. Richards, *J. Chem. Soc. Dalton* (1981) 1629–1634.

(35) D. S. Edwards, L. V. Biondi, J. W. Ziller, M. R. Churchill, R. R. Schrock, *Organometallics* **2** (1983) 1505–1513.

(36) M. R. Churchill, H. J. Wasserman, H. W. Turner, R. R. Schrock, *J. Am. Chem. Soc.* **104** (1982) 1710–1716.

(37) M. R. Churchill, W. J. Youngs, *Inorg. Chem.* **18** (1979) 171–175.

(38) D. Neugebauer, E. O. Fischer, N. Q. Dao, U. Schubert, *J. Organomet. Chem.* **153** (1978) C41–C44.

(39) E. O. Fischer, F. J. Gammel, D. Neugebauer, *Chem. Ber.* **113** (1980) 1010–1019.

(40) E. O. Fischer, P. Friedrich, T. J. Lindner, D. Neugebauer, F. R. Kreißl, W. Uedelhoven, N. Q. Dao, G. Huttner, *J. Organomet. Chem.* **247** (1983) 239–246.

(41) E. O. Fischer, W. Röll, N. H. Tran-Huy, K. Ackermann, *Chem. Ber.* **115** (1982) 2951–2964.

(42) S. J. Holmes, R. R. Schrock, M. R. Churchill, H. J. Wasserman, *Organometallics* **3** (1984) 476–484.

(43) F. A. Cotton, W. Schwotzer, *Inorg. Chem.* **22** (1983) 387–390.

(44) A. Mayr, A. M. Dorries, G. A. McDermott, S. J. Geib, A. L. Rheingold, *J. Am. Chem. Soc.* **107** (1985) 7775–7776.

(45) M. R. Churchill, A. L. Rheingold, H. J. Wasserman, *Inorg. Chem.* **20** (1981) 3392–3399.

(46) K.W. Chiu, R. A. Jones, G. Wilkinson, A. M. R. Galas, M. B. Hursthouse, K. M. A. Malik, *J. Chem. Soc. Dalton* (1981) 1204–1211.

(47) E. O. Fischer, D. Wittmann, D. Himmelreich, U. Schubert, K. Ackermann, *Chem. Ber.* **115** (1982) 3141–3151.

(48) A. C. Filippou, E. O. Fischer, H. G. Alt, U. Thewalt, *J. Organomet. Chem.* **326** (1987) 59–81.

(49) H. Fischer, F. Seitz, J. Riede, *J. Chem. Soc. Chem. Comm.* (1985) 537–539.

(50) W.W. Greaves, R. J. Angelici, B. J. Helland, R. J. Klima, R. A. Jacobson, *J. Am. Chem. Soc.* **101** (1979) 7618–7620.

(51) E. O. Fischer, H. Hollfelder, P. Friedrich, F. R. Kreißl, G. Huttner, *Angew. Chem.* **89** 416–417; *Angew. Chem. Int. Ed. Engl.* **16** (1977) 401–402.

(52) E. O. Fischer, T. L. Lindner, G. Huttner, P. Friedrich, F. R. Kreißl, J. O. Besenhard, *Chem Ber.* **110** (1977) 3397–3404.

(53) M. R. Churchill, Y.-J. Li, L. Blum, R. R. Schrock, *Organometallics* **3** (1984) 109–113.

(54) M. Green, J. A. K. Howard, A. P. James, A. N. de Jelfs, C. M. Nunn, F. G. A. Stone, *J. Chem. Soc. Chem. Comm.* (1984) 1623–1625.

(55) M. R. Churchill, W. J. Youngs, *Inorg. Chem.* **18** (1979) 2454–2458.

(56) F. A. Cotton, W. Schwotzer, E. S. Shamshoum, *Organometallics* **3** (1984) 1770–1771.

(57) S. L. Latesky, J. P. Selegue, *J. Am. Chem. Soc.* **109** (1987) 4731–4734.

(58) M. H. Chisholm, D. M. Hoffman, J. C. Huffman, *Inorg. Chem.* **22** (1983) 2903–2906.

(59) M. H. Chisholm, B. K. Conroy, J. C. Huffman, *Organometallics* **5** (1986) 2384–2386.

(60) M. H. Chisholm, J. C. Huffman, N. S. Marchant, *J. Am. Chem. Soc.* **105** (1983) 6162–6163.

(61) S. M. Rocklage, R. R. Schrock, M. R. Churchill, H. J. Wasserman, *Organometallics* **1** (1982) 1332–1338.

(62) M. R. Churchill, Y.-J. Li, *J. Organomet. Chem.* **282** (1985) 239–246.

(63) F. A. Cotton, R. A. Walton, *Multiple Bonds Between Metal Atoms*, New York: Wiley, (1982).

(64) M. D. Curtis, W. M. Butler, *J. Organomet. Chem.* **155** (1978) 131–145.

(65) U. Schubert, in *Transition Metal Carbene Complexes,* Weinheim: Verlag Chemie, (1983), pp. 71–111. U. Schubert, *Coord. Chem. Rev.* **55** (1984) 261–286.

(66) U. Schubert, *Organometallics* **1** (1982) 1085–1088.

(67) A. A. Westenberg, J. H. Goldstein, E. B. Wilson, *J. Chem. Phys.* **17** (1949) 1319–1321. C. C. Costain, *J. Chem. Phys.* **23** (1955) 2037–2041.

(68) N. M. Kostić, R. F. Fenske, *J. Am. Chem. Soc.* **103** (1981) 4677–4685.

(69) D. Saddei, H. J. Freund, G. Hohlneicher, *J. Organomet. Chem.* **216** (1981) 235–243.

(70) B. E. R. Schilling, R. Hoffmann, J.W. Faller, *J. Am. Chem. Soc.* **101** (1979) 592–598.

(71) A. R. Rossi, R. Hoffmann, *Inorg. Chem.* **14** (1975) 365–374.

(72) N. M. Kostić, R. F. Fenske, *Organometallics,* **1** (1982) 489–496.

Electronic Structures of Transition Metal Carbyne Complexes

By Peter Hofmann

1 Introduction

Fifteen years ago a new class of organometallic compounds appeared on the stage of contemporary chemistry. The pioneering work of E. O. Fischer and his collaborators yielded the first mononuclear transition metal carbyne complexes, exemplified by **1** as a typical example of the first group of such compounds with transition metal carbon triple bonds (1).

$$Br-W\equiv C-CH_3 \qquad\qquad L_nM\equiv C-R$$

$$\mathbf{1} \qquad\qquad\qquad \mathbf{2}$$

The accessibility of **1** and of related derivatives ultimately evolved from the Fischer group's detailed and persistent investigation of transition metal carbene complexes and their reactivity patterns. The discovery of carbyne complexes has opened a large and still rapidly expanding area of organometallic research, comparable to the situation in the field of carbene and alkylidene transition metal systems, which had been introduced by E. O. Fischer and his coworkers nearly a decade earlier (2). The most numerous and important contributions to our present knowledge about the chemistry of carbyne complexes originate from Fischer's group, but during the past fifteen years many other researchers have been equally active participants in this subfield of metal-to-carbon multiple bond chemistry. The number of known transition metal carbyne (alkylidyne) compounds, which have been fully characterized, today probably exceeds 300. A couple of review articles (3) on the subject have appeared, the most recent extensive one in 1987 (4). Five years after the first report on a carbyne complex by Fischer et al., in 1978, Schrock (3e) and his collaborators introduced a second novel type of carbyne (alkylidyne) complexes. In contrast to the "Fischer complexes", these molecules were described as containing transition metal centers (predominantly Mo and W) in (at least formally) high oxidation states. Typical examples are tetracoordinate $W(CBu^t)(CH_2Bu^t)_3$ and $W(CR)(OBu^t)_3$ or the fascinating pentacoordinate molecule $W(CBu^t)(CHBu^t)(CH_2Bu^t)(Me_2PCH_2\text{-}CH_2PMe_2)$, in which an intramolecular comparison of alkyl, alkylidene and alkylidyne bonding to the common metal center is possible. The assignment of high (VI) oxidation states to the metals in these "Schrock-type" alkylidyne complexes is the outcome of an electron counting formalism which, to a large extent, is a matter of taste. This convention does not necessarily portray realistic details of electronic structures and electron density distributions. We come back to this point later.

Historically it is interesting to note, how the development of Schrock's transition metal alkylidyne systems broadened and fertilized the field of metal carbon triple bond chemistry to a similar extent as had already occurred with "Schrock-type" alkylidene complexes and their older "Fischer-type" carbene complex relatives.

Whereas at first carbyne and alkylidyne complexes seemed to be somewhat exotic and of pure academic interest, the last couple of years have unravelled more and more fascinating facets of the chemistry of these organometallic species. There is growing interest in the reaction chemistry of carbyne and alkylidyne complexes, highlighted for instance by Schrock's discovery of tungsten(IV)-alkylidyne mediated acetylene metathesis (5), or by carbon-carbon coupling reactions yielding organic molecules within the coordination sphere of a transition metal (6).

Despite the increasing tendency to employ transition metal carbyne complexes for synthetic purposes in various areas of organometallic or organic chemistry, or to explore their propensity to function as catalysts, less attention has been devoted to their physical properties and theoretical chemistry, compared to, for example, their carbene and alkylidene congeners. There exists a respectable body of structural work, ranging from standard routine X-ray investigations to high quality determinations of electron deformation densities in carbyne complexes (*vide infra*). These data have given much insight into structural details. On the other hand the number of serious attempts to elucidate and explore the electronic structural characteristics of transition metal compounds represented by the general formula 2, either by spectroscopy (other than routine type techniques) or by quantum chemical electronic structure calculations, is rather limited.

This review summarizes work published to date dealing with questions of bonding and electronic structures of transition metal carbyne complexes. The discussion exclusively deals with mononuclear systems and follows the original definition of a carbyne complex as given by Fischer (3a). Here carbyne complexes are defined as compounds where: (i) a carbyne group (CR) is bound to a single metal, (ii) the carbyne carbon is largely sp-hybridized, and (iii) the metal carbon bond possesses triple bond character or is at least significantly shorter than a double bond. This definition, although mainly operational and rather formal, encloses all typical "Fischer-type" carbyne and "Schrock-type" alkylidyne complexes. Not included are molecules with carbyne units bridging two or more transition metal centers, because, in an electronic sense, these compounds are quite different from mononuclear species. Carbynes in μ_2- or μ_3-bridging situations are known for many di- or trinuclear metal arrangements. The extensive work of Seyferth on alkylidynetricobalt nonacarbonyl clusters (7) and of Stone (8), who prepared an enormous variety of beautiful μ_2- and μ_3-carbyne complexes with different numbers and types of transition metal centers, following the guidelines of the Isolobal Analogy (9), exemplify this chemistry of CR-groups bonded to metals in an alternative way.

While experimental chemistry of carbyne complexes has rapidly and continuously expanded ever since their first synthesis in 1973, electronic structure calculations did not begin to appear until around 1980. At that time, a sufficient number of compounds **2** with metal carbon triple bonds had been studied by X-ray diffraction, vibrational spectroscopic (10) or NMR methods, and the first alkylidyne complexes had been prepared and fully characterized. At present approximately a dozen theoretical papers have appeared in the literature, which adress bonding and electronic structure problems of transition metal carbyne complexes, employing quantum chemical calculations of different degrees of sophistication. The methodologies utilized range from semiempirical approaches to correlated *ab initio* computations beyond the Hartree-Fock limit. We will focus upon various aspects of these studies in the course of this chapter. A rather consistent and satisfactory picture emerges from the available body of theoretical investigations, as long as we are aiming towards a general, qualitatively reliable, transferable description and understanding of the bonding and molecular orbital structure of carbyne complexes. The simplified term "molecular orbital structure" is used here, because all available theoretical investigations have made use of molecular orbital (MO) theory as the basic underlying formalism for treating electronic structures of carbyne complexes. To the author's knowledge neither VB-type nor $X\alpha$ or related density function theoretical computations have been published for the class of organometallic compounds under consideration here.

It is the intention of this review to describe and to discuss the main features and conclusions of the various theoretical treatments of carbyne complexes, as they have appeared in the literature. It seems reasonable, however, to begin with a general and – as far as possible – method independent bonding picture for prototype molecules. To do so we will make use of the fragment orbital approach, which allows us to "construct" the valence molecular orbitals of a typical transition metal carbyne complex from either well known or easily conceivable valence MOs of appropriate building blocks (11). This approach of relating molecular orbitals of a molecule back to those of its constituent fragments, by constructing an appropriate interaction diagram, is an efficient tool in electronic structure theory of qualitative and quantitative nature alike. The basic electronic structure of "Fischer-type" transition metal carbene complexes has been interpreted by using the fragment MO method (12). The majority of molecular orbital studies of carbyne complexes to be adressed below also employs fragment MO considerations or explicit fragment MO calculations (basis set transformations from atomic to molecular fragment basis sets) as an interpretational aid to rationalize or to explain bonding mechanisms and numerically computed electronic details.

A "natural" dissection of transition metal carbyne (or alkylidyne) complexes into suitable fragments leads to a description of systems **2** as being composed of a transition metal-ligand subunit ML_n and a carbyne group CR. Let us turn to the less complicated building block, the CR moiety, first.

2 Carbyne Ligands

At this point it is informative to digress briefly from the mere qualitative description of simple carbyne ligand systems, as it is to be found in the various reviews cited above (3, 4). Carbynes CR, and in particular the parent diatomic CH, **3**, represent electronically and coordinatively highly unsaturated and reactive species. CH itself has been well characterized in the gas phase, its electronic ground state has been determined (13).

$$:\!C\!-\!H \qquad \bigcirc C\!-\!H \qquad \underset{\bigcirc}{\overset{\bigcirc}{C}}\!-\!H \qquad \ominus\!-\!H \qquad \underset{\bigcirc}{\overset{\bigcirc}{C}}\!-\!H$$

$$\textbf{3} \qquad\quad \textbf{4}\ (\sigma) \qquad \textbf{5}\ (p_x) \qquad \textbf{6}\ (p_y) \qquad \textbf{7}$$

Simple one-electron MO theory tells us that one of the four carbon electrons is engaged in the CH σ-bond, ie, in a low lying, doubly occupied molecular orbital, and that we are left with three orbitals for the remaining 3 electrons of carbon.

These 3 valence orbitals may be described as canonical MOs, **4–6**, or, alternatively, as 3 equivalent hybrid orbitals shown in **7**. Different states can be constructed for different occupation and spin patterns of **4–6** or **7**. As will become evident later, we do not have to consider these states in detail, if we want to discuss the electronic structure of carbyne complexes. But it is rather informative with respect to an eventual evaluation of quantitative MO studies, to be addressed later in the course of this account, to take a brief look at the electronic structure of the simple diatomic CH in order to compare experiment and quantitative electronic structure calculations. Experimentally the electronic ground-state of CH is a $^2\Pi$ state (formally 2 electrons in **4**, 1 electron in **5** or **6**, in the simple one-electron MO picture). At least 71 kJ/mol above the groundstate is found a $^4\Sigma^-$ state (3 unpaired electrons). The relevant point to note here, and the lesson to be learned for later, is that even an approximate quantitative reproduction of the correct state energy ordering for a fragment as simple as CH requires a level of quantum theory which, at present, is absolutely out of reach for transition metal systems like carbyne complexes. Calculations for CH at the HF/6-31G** level incorrectly make the $^4\Sigma^-$ state more stable than $^2\Pi$ by about 29 kJ/mol (14), and taking into account electron correlation up to 3rd and 4th order using the same basis set (MP4/6-31G**) still gives a state separation error of around 42 kJ/mol. It is necessary to go to the MP4/6-311G**//6-31G*/or a comparable level (14, 15) in order to reproduce reasonably well the energy separation between the ground ($^2\Pi$) and first excited ($^4\Sigma^-$) states of CH (79.9 kJ/mol calculated versus 71 kJ/mol experimentally). Certainly the task of performing numerically accurate and reliably predictive electronic structure calculations for state energies, dissociation energies etc. of an open shell system like CH falls into a different category compared to a quantum chemical treatment of the basic bonding pattern of stable, closed shell transition metal carbyne complexes. Nevertheless, the level of sophistication needed for the primitive diatomic CH sheds some light upon the significance and limitations of carbyne complex MO results to be addressed later.

For our intended qualitative bonding description of prototype metal carbyne complexes the "true" state ordering of CH or CR groups as such is irrelevant. If we construct a molecule of type **2** by conceptually uniting a metal fragment ML_n and an alkylidyne subunit, the distribution of valence electrons among the two fragments is more conveniently chosen by following the usual habit (9) of beginning with building blocks that do not hold unpaired electrons. CH or CR ligands will be treated as formally cationic 2 electron donors, making them similar to NO^+. A cationic CH^+ ligand then provides a doubly filled donor orbital of σ-symmetry, **4**, and two degenerate, orthogonal, low lying acceptor MOs of π-symmetry, **5** and **6**, corresponding to the π* levels of NO^+. It should be emphasized, that this choice of counting carbyne ligands as cationic two electron donors is purely formalistic, and that an alternative description as CR^{3-} is equally well justified. Neither way of allocating electrons and assigning oxidation states to the metal or to the carbyne ligand is "better" or "more justified" – it is the electron distribution of the composite molecule that counts. So let us examine some typical carbyne ligands using the CR^+ convention.

For CH^+, molecular orbital **4**, the lonepair, is largely localized on carbon, orbitals **5** and **6** are pure 2p AOs of carbon and appear at very low energy. A CH^+ ligand therefore on energy and overlap grounds is a good σ-donor and an extraordinarily capable, double faced, π-acceptor. This picture is of course altered, if the hydrogen atom of CH^+ is replaced by varying substituents R, and the extent of changes brought about by going from CH^+ to, say, $C(CH_3)^+$ and $C(NH_2)^+$, **8** and **9**, is easily understood.

Practically all known and well characterized carbyne complexes contain CR ligands, R being alkyl, aryl, silyl, alkoxy, siloxy, amino, thioalkyl etc.; the only exception, bearing a simple methylidyne ligand in a mononuclear system is found in a few compounds of a type represented by **10** (16).

$$Cl—W\equiv CH \quad \textbf{10}$$

Figure 1 shows a comparison of the valence orbitals for CH^+, **8** and **9**, as they emerge from a simple extended Hückel calculation. A short glossary upon this largely self-explaining diagram is useful, because these three types of alkylidyne ligands are the ones that play the main role in the theoretical studies to be considered at a later point. Furthermore, aminocarbyne as well as the less common alkoxycarbyne ligands in a sense represent an extreme end of the scale, if we want to discuss carbyne carbon substituents with respect to their ability to influence the bonding of the carbyne fragment to the metal side.

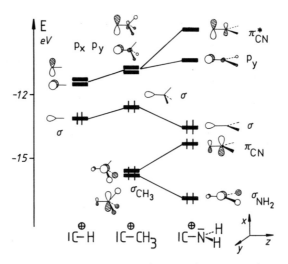

Fig. 1: Comparison of the valence orbitals of CH^+, $C(CH_3)^+$ and $C(NH_2)^+$ derived from extended Hückel calculations.

Comparison of CH^+ with the simplest alkyl derivative $C(CH_3)^+$ in *Figure 1* reveals one consequence of replacing H by CH_3, namely a slight destabilization of the (still degenerate, C_{3v}-symmetry) acceptor level set (p_x, p_y). Due to an antibonding admixture of a small amount of CH_3 group orbitals of appropriate symmetry (the C-H σ-bonding methyl MOs with π-character) into p_x and p_y, these acceptor levels now show a slight delocalization onto the CH_3 part of the fragment. At much lower energy the CH_3 σ-bonding group orbitals (σ_{CH_3}) themselves are found. According to the rules of standard perturbation theory (11) these MOs of $C(CH_3)^+$ as they are shown in *Figure 1*, are simply the MO equivalents of hyperconjugation between the carbyne carbon and the methyl group. Overall, $C(CH_3)^+$ and other alkylcarbyne ligands are electronically not very different from CH^+. Alkyl groups may provide important stabilizing factors in transition metal carbyne chemistry, however, owing to a combination of various effects, like steric shielding, σ-donation, charge delocalization etc., depending on the specific system.

In aryl substituted carbyne ligands, eg, $C(C_6H_5)^+$, phenylcarbyne, only one of the p_x, p_y acceptor AOs of the carbyne center can interact with the aryl π-system, and the degeneracy of the acceptor level set is lifted to a certain degree. We shall not dwell upon arylcarbyne systems here, because phenylcarbyne complexes are mentioned later. The phenomenon of breaking the twofold degeneracy of the carbyne ligand acceptor levels p_x and p_y is strongly pronounced, however, in aminocarbyne and oxycarbyne ligands, where the p-type lonepairs of the adjacent nitrogen or oxygen atom, interacting with either p_x or/and p_y, lead to a significant contribution of resonance structure **9b** for, eg, $C(NH_2)^+$. *Figure 1* shows, that going from CH^+ or $C(CH_3)^+$ to $C(NH_2)^+$ leads to inductive stabilization of the σ-donor MO and to an extensive energy gap between p_y and π^*_{CN}, the two acceptor levels of the aminocarbyne fragment (17). Essentially aminocarbyne ligands possess only one acceptor MO (p_y) comparable in energy to those of CH^+ or CR^+. The other one, at higher energy, now is the π^* orbital of the C-N double bond. Its bonding counterpart π_{CN} appears somewhat below the σ-donor level of $C(NH_2)^+$ and is to be correlated with the CH_3 bond orbitals of $C(CH_3)^+$, as is the σ-MO of the NH_2 group, σ_{NH2}. Valence MOs π^*_{CN}, p_y, σ and π_{CN} will determine the interactions of an aminocarbyne ligand with a metal fragment (18).

Aminocarbynes consequently are closely related to vinylidene ligands (viz the azavinylidene resonance structure **9b**), and to a large extent they are approaching the "single faced" acceptor ligand type with *one* dominating acceptor MO lying in the ligand plane. A $C(NH_2)^+$ or $C(NR_2)^+$ fragment should, however, be a better π-acceptor than a vinylidene (π^* of vinylidene lies above π^*_{CN}) or a carbonyl group, where *both* π^* acceptor levels approximately correspond to π^*_{CN} of $C(NH_2)^+$. The distinct energetic (and overlap based (19)) differentiation between p_y and π^*_{CN} will naturally lead to a rotational barrier and to preferred rotational orientations of the aminocarbyne plane for metal fragments ML_n without rotationally symmetric donor orbital sets *(vide infra)*.

It is easy to derive qualitatively or to compute valence orbital pictures analogous to *Figure 1* for other carbyne ligands CR^+. The three prototypes described above suffice for our needs in the remaining body of this review, however.

We turn now to some general considerations about metal fragments ML_n, in order to be prepared for treating the bonding in carbyne complexes and for a survey of the available theoretical work.

3 Metal Fragments in Mononuclear Carbyne Complexes – General Bonding Aspects

The valence orbital patterns and bonding capabilities of CR^+ fragments, discussed in the previous section, place certain demands upon a metal-ligand fragment ML_n for it to qualify as a bonding partner of a CH^+ or CR^+ moiety. Again we may choose *a priori* whatever electron distribution we prefer for the alkylidyne or ML_n moiety, respectively. We just have to be consistent in our choice in surveying metal fragments. Accordingly, we retain our formalism whereby carbyne fragments are taken as cationic, 2 electron donors/4 electron acceptors, in direct analogy with the view of a nitrosyl ligand as NO^+. The electronic similarity of both groups is evident and has been stressed many times. Assigning the alkylidyne ligands this way predetermines the ML_n charge as ML_n^- for neutral, ML_n for cationic, and ML_n^{2-} for anionic carbyne complexes. From the usual way of electron counting, taking ligands as 2, 4, 6 etc. electron donors (9, 11) the overall d electron count of M in an ML_n fragments is then uniquely defined.

Considering only closed shell transition metal carbyne complexes, either cationic, neutral or anionic (20), many prototypical metal fragments with an even number of d electrons may in principle lead to a stable triple bonding situation between ML_n and a carbyne unit. By "prototypical" metal fragments we refer to the d^m–ML_n classification as developed within the Isolobal Analogy (9). It only concentrates upon d-electron count, number of ligands L and the specific coordination geometry of ML_n units, without specifying M and L in detail. In order to provide triple bonding capability towards a CR^+ ligand, certain basic prerequisites of ML_n are mandatory. First, ML_n units must be coordinatively unsaturated, which by definition makes at least one σ acceptor MO at the metal available and requires 16 or fewer valence electrons, in order not to exceed an 18 electron count when the CR^+ fragment is attached. Furthermore, a minimum d electron count of d^4 is necessary, and these four d electrons have to be in ML_n molecular orbitals of π-symmetry oriented towards the available coordination site of the carbyne. For transition metal fragments such backbonding MOs typically are d-AO based levels or some sort of dp-hybrids. Moreover, in order that the two π-bonding interactions (d to p_x and p_y) for an MC triple bond are not only possible but also efficient, the two ML_n fragment MOs housing the 4 electrons ought to be degenerate or at least close in energy, since p_x and p_y of most CR^+ ligands are practically degenerate. Finally, these two relevant metal levels should have a low ionization energy, readily available as high lying molecular orbitals of ML_n, rather than being strongly stabilized by, eg, a set of acceptor ligands, and be properly hybridized and of suitable spatial extension to ensure good overlap with the CR^+ ligand. All but the last two of these conditions are mainly a matter of geometry, symmetry and d electron count; the latter requirements depend upon the specific ML_n fragment in each case.

Among the most widely used metal fragments are d^6-ML_5 units. The basic orbital structure of such octahedral fragments has been discussed many times in the literature ((9, 11, 12) and references therein). In **11** the four relevant valence MOs of d^6-ML_5 (C_{4v}) are sketched. MOs a_1, e_s and e_a are the ones available for bonding to a CR^+ unit: a_1 interacts with σ of CR^+ (viz. *Figure 1*), e_s and e_a, a degenerate set of two d-type MOs, interact with p_x and p_y. An MC triple bond results.

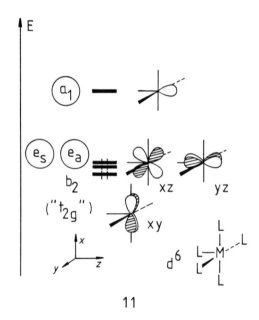

11

Many actual metal fragments belong to this d^6-ML_5 category. Several typical examples, occurring in various carbyne complexes, are shown in **12–16** with their appropriate ML_n charge. Many others, more complicated and often less symmetrical ones are found in known compounds.

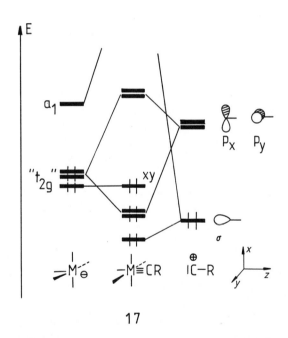

14   15   16

In **12**, **13** and **16** true e-sets of donor levels as depicted in **11** are present; **14** and **15** also provide two orthogonal π-donor MOs of nearly equal energy (21). In all these cases a set of three MOs, a remnant of the original octahedral ML_6 t_{2g} set with 6 electrons, is present, and two of these orbitals with their 4 electrons can engage in π-bonding to a carbyne carbon atom. This interaction is qualitatively shown in **17** and results in a triple bond, leaving one of the metal levels (xy) unaffected.

17

From **17** one notes that for an electron count of d^4 a low lying empty d level and consequently a small HOMO-LUMO gap would be present. Such a case would still allow the triple bond to exist, but would render electronic instability to the system, unless this remaining d level is pushed up in energy by some appropriate mechanism. This is exactly what happens in d^4-ML_5 based transition metal

carbyne complexes. A variety of these is known, two typical examples from
Schrock's laboratory are depicted in **18** and **19** (22, 23a), both having been
studied by X-ray structural analysis.

In both **18** and **19** π-donor ligands interact with the unfilled d-level (xy) resulting from level scheme **17** for a d^4 metal center, pushing it to higher energy. This
interaction secures an electronically stable structure for a carbyne complex of a
d^4-ML_5 unit. The alternative convention of calling systems like **18** and **19** W(VI)
complexes, counting the carbyne ligands as CR^{3-}, also leads to the qualitative
picture of **17** with 6 electrons in the MC triple bond MOs and one remaining
"empty" d-level, which interacts with the π-donors, consequently gains electron
density and is pushed up, thereby diminishing the compounds electrophilicity
at the metal. Obviously this description is just the MO equivalent of saying that
a high oxidation state is stabilized by a specific ligand set of π-donors.

In passing we note, that the X-ray structure of a seven-coordinate tungsten
alkylidyne complex with a rare regular capped octahedral geometry has also
been determined (23b); here a d^4-ML_6 metal fragment $(PMe_3)_3Cl_3W^-$ binds to a
CBu^{t+} ligand. We will not deal with d^4-ML_6 units and seven-coordinate carbyne
complexes in detail here, however, as lower coordination numbers are much
more common.

Another class of metal fragments found in stable carbyne complexes belongs
to the d^8-ML_4 category.

Roper has prepared (3d) a series of trigonal bipyramidal Os-carbyne systems
with the carbyne ligand in an equatorial position, as expected on electronic
grounds (24a) for a good acceptor. **20** displays the valence MO pattern for a
general d^8-ML_4 fragment of C_{2v} symmetry, eg, $Fe(CO)_4$ or $Re(CO)_4^-$. Here the
two levels appropriate for π-bonding to CR^+, b_2 and b_1, are nondegenerate, but
interactions both with p_x and p_y of carbyne groups are still possible.

21 represents such a building block d^8-ML_4, typically found in Roper's carbyne complexes (3d) and also in a cationic aminocarbyne complex, (PPh_3)-
$(CO)_3FeCN(Pr^i)_2^+$, made by Fischer et al. (24b). It is interesting to note how **21**,
owing to its specific ligand set, is again tailored in such a way, as to make it elec-

tronically much better suited for bonding to CR^+ than, say, an isoelectronic and isolobal $Re(CO)_4^-$ would be. Both the axial phosphine ligands and the equatorial chlorine raise the b_1 MO in **20** to higher energy, because axial PR_3 groups (unlike CO) cannot push this level down by strong backbonding. Moreover the lonepair of Cl destabilizes b_1 by mixing into this MO with antibonding character "from below", as indicated in **22**. Halogen ligands situated trans to the carbyne generally play an important stabilizing role in many octahedral "Fischer-type" transition metal carbyne complexes (see also **13** and **16**), as will become apparent later. For the iron system of Fischer et al. (24b), $(PPh_3)(CO)_3FeCN(Pr^i)_2^+$, the amino group partially satisfies the electron demand of the carbyne carbon, and the axial PR_3 ligand again raises the metal fragment b_1 level (as in Roper's systems) compared to its $Fe(CO)_4$ parent. The result is overall stabilization. A second class of d^8-ML_n fragments, namely T-shaped d^8-ML_3, capable of binding carbyne units, is related to d^6-ML_5 (**9**) and will not be discussed further here.

Typical representatives of Schrock's alkylidyne complexes are of type L_3MCR, exemplified by $(CH_2Bu^t)_3W(CBu^t)$ or $(OBu^t)_3W(CBu^t)$. In our counting convention these systems contain d^4-ML_3 metal fragments, (R_3W^- or $(RO)_3W^-$), bound to $C(Bu^t)^+$. The X-ray structure determination of **23** (25) reveals a nearly tetrahedral, monomeric molecule in the crystal. Formally **23** is a 12 electron system, if we count alkoxy ligands as 2 electron donors, but of course their π-donating capability makes such a formalism artificial in any counting scheme, be it the M(VI) alternative or our choice of using an $(RO)_3W^-/CR^+$ electron count.

23

Because the section of this article, which later surveys published theoretical work on transition metal carbyne molecules does not encompass L_3MCR complexes of the type just mentioned, we briefly discuss their electronic structure qualitatively here, on the basis of what is known about ML_3 fragments. Interestingly, to our knowledge, no detailed quantum chemical study of L_3MCR carbyne complexes has been published. This situation is surprising, because not only are such complexes the structurally least complicated representatives of this class of molecules, but they also are of great interest because of their potential in alkyne metathesis. A GVB study (26) of their "cycloaddition products" with alkynes, $L_3MC_3R_3$, has appeared very recently, but without reference to calculations for L_3MCR molecules.

The electronic and molecular orbital structures of the metal fragments in L_3MCR carbyne species, which, according to the counting procedure used here, are of the type d^4-ML_3, (eg $(RO)_3W^-$) is well known. ML_3 fragments are extremely common building blocks in inorganic and organometallic chemistry, and their generalized MO picture and bonding capabilities have been frequently analyzed and described in the literature (27) for different electron counts and various ligands L (28). Usually the valence orbitals of an ML_3 (C_{3v}) fragment are thought of as being derived from those of octahedral ML_6 (11) by formally stripping off three facial ligands. For the purpose of discussing essentially tetrahedral L_3MCR molecules here, it is important to note that the L-M-L angles are around 110°, ie, larger than in an octahedral fragment (90°). For such "tetra-hedral" ML_3 units the AO composition of the two e orbital pairs 1e and 2e is as sketched in **24** (27). Orbital 1e is composed mainly of xz and yz character, while 2e is mostly of xy and x^2-y^2 type. The "tilting" shown for 1e and 2e in **24** is a consequence of the mixing of the two sets of symmetry equivalent d-orbitals. MOs $1a_1$ and $2a_1$ are the z^2 and an sp hybrid MO (29).

24

The AO character, shown in **24** is also the one preferred for π-donor ligands L in ML_3, in contrast to pure σ-donor or π-acceptor ligand sets, where 1e tends to have more x^2-y^2 and xy components and 2e is predominantely xz and yz. So for fragments like $(RO)_3M^-$, with "tetrahedral" L-M-L angles the simplified picture of **24** applies. A CH^+ or CR^+ moiety can therefore nicely interact with such an ML_3^- unit and **25** presents a qualitative interaction scheme, which goes back to EH model calculations (30).

The three occupied molecular orbitals in **25** represent the triple bond between the metal and the carbyne ligand; there is an appreciable gap between these MOs and the next three empty levels at higher energy. When a fourth ligand, a CR^+ in this case, restores the tetrahedron, as in **25**, two three-orbital interactions (p_x and p_y with 1e and 2e, and σ with $1a_1$ and $2a_1$) are turned on. Following the rules for orbital interactions this bonding leads to the outcome represented in **25**: three empty d levels (z^2, xy, x^2-y^2, nearly pure d AO's) appear in between the

25

π and π^* set of the MC triple bond. The π-donor ligands make them evolve at rather high orbital energy, and together with an appropriate amount of steric shielding (by, eg, the OBu^t or similar space filling donor ligands), stabilize such formal 12 electron systems even as monomers. Scheme **25**, however, also explains the tendency of certain $(RO)_3MCR$ systems to form alkoxy bridged dimers or to bind additional ligands. The 3 empty d levels function as potent acceptor orbitals and, if steric limitations permit, binuclear structures such as **26** are formed (31), or donor ligands like pyridine coordinate to the metal. The mainly z^2 based empty MO depicted in **27** facilitates the bonding interaction to an oxygen lonepair of a second molecule in **26**, or with other ligands. A fifth ligand, binding as shown in **27**, changes the metal coordination geometry towards trigonal bipyramidal, which in turn also raises the empty e-set (xy, x^2-y^2) of scheme **25** through stronger interactions with the OR-group lonepairs, now in an equatorial position.

There is a second type of ML_3 fragments, which according to the previously stated general requirements, could potentially bind to a CR^+ ligand, namely, pyramidal (C_{3v}) d^{10}-ML_3. This possibility is already evident from **25**. For a d^{10}

26   27

electron count the three metal orbitals of L_3MCR in the center would be filled. For a d^{10}-ML_3 unit, the valence MO splitting pattern ($1a_1$, $1e$, $2e$, $2a_1$) of **24** still applies. Only the relative percentage of xz and yz character in 2e (and of x^2-y^2, xy in 1e) is larger for typical ligands found in complexes having d^{10}-ML_3 subunits (phosphines, CO etc.). To our knowledge no neutral carbyne complexes of type **28** or isoelectronic analogs are known to date, nor has any system **29**, an equivalent of the well known CpNiNO, been made. A cationic aminocarbyne nickel complex, related to systems **28** is known, however (47b).

28   *(M=Co,Rh,Ir)*   29

Of course simple isolobal relationships do not reveal true thermodynamic or kinetic stabilities of specific molecules. In order to evaluate "isolobally" acceptable fragment combinations, such as those in **28** or **29**, at least semiquantitative model calculations would be necessary, but are not available at present. There are several reasons to expect low kinetic stability and high reactivity for species like **28** and **29**, including an energetically high lying MC σ-bond, lability of the M-P bonds, and insufficiently strong MC π-bonding, but nevertheless a detailed experimental search and theoretical exploration of such systems ought to be worthwhile. Only a single mentioning of the possible intermediacy of neutral d^{10} ML_3CR complexes together with some EH-calculations, has appeared in the literature (see chapter 4).

In closing this chapter, let us turn to a final prototype representative of transition metal fragments. One of the widely employed ML_n units in organotransition metal chemistry is a bent Cp_2M sandwich fragment. For M = V, Nb and Ta of formal oxidation state (I), such bent Cp$_2$M fragments or their alkyl substituted analogs would carry a negative overall charge. They are of the required d^4 electronic configuration to allow, at least in principle, a triple bond to CH$^+$ or CR$^+$ ligands. Since the fascinating molecule **30**, with a vanadium to nitrogen triple bond, has been recently prepared and its structure determined (32), it seems reasonable to discuss briefly the qualitative electronic structure for molecules like **31** (33).

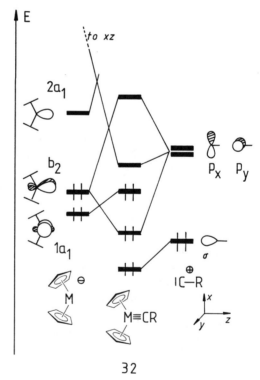

The valence MOs of a bent Cp$_2$M unit, available from various sources (34), are given on the left side of scheme **32**, in which is sketched a qualitative interaction diagram for molecules like **31**.

This diagram reveals that a d^4-Cp_2M metal fragment, owing to the nature of its three low lying valence MOs, can only turn on *one* strong π-bonding interaction to the acceptor level set of the carbyne group (to p_y). Two of the four metal electrons stay in a metal level ($1a_1$, y^2) that is somewhat destabilized in the composite molecule because of its repulsion with σ of CR^+. The xz MO of Cp_2M, however, needed for the second π-bond to p_x of the carbyne ligand, lies quite high in energy (not shown explicitly in **32**), being pushed up by the pair of Cp ligands. An electronically unstable situation results (35). One strategy to overcome this situation, would be to raise the LUMO in scheme **32** by going to aminocarbyne ligands (see above), and indeed a cation $Cp_2WCNEt_2^+$, isoelectronic to **31**, seems to be isolable (36). The stability of **30** derives from the much lower energy of NR^+ levels p_x and p_y compared to CR^+, owing to the greater electronegativity of nitrogen. The electronic structure of **30** which has been discussed (32), can be derived from **32**. For Cp_2VNPh, a model for **30** employed in Xα calculations (32), the metal to nitrogen π-bonding level ($9b_2$ in (32)) derived from p_x and xz on V, only shows 17 % vanadium character as well. Contrastingly, the other π-orbital, coming from b_2 and p_y in **32**, is localized on vanadium with more than 50 % of the corresponding wavefunction, in accord with **32**, modified by replacing CR^+ with NR^+.

Having discussed several, but by no means all possible, typical metal fragments and some general aspects of metal-to-carbyne bonding in a merely qualitative way, we now turn to the results of actual electronic structure calculations for mononuclear transition metal carbyne and alkylidyne complexes, as they have appeared in the literature.

4 Electronic Structure Calculations

This section is organized in the simplest possible way by discussing available published work in strictly chronological order. In toto, the number of electronic structure computations for mononuclear transition metal carbyne complexes comprises fewer than about a dozen papers. Strangely enough, it took around 7 to 8 years from the discovery of Fischer's first carbyne complexes, before theoreticians began to take a closer look at this class of compounds. Moreover, as stated above, to the best of the author's knowledge "Schrock-type" systems like **18**, **19**, **23** or others, discovered around 10 years ago, still have not been studied computationally.

The first paper that turned any attention to the bonding situation in carbyne systems, came from R. Hoffmann and coworkers (37). This analysis was carried out in the context of analyzing unusual M-C-H angles, C-H bond activation and α-hydrogen abstraction in transition metal carbene complexes. For a highly simplified Ta(I) d^4 transition metal carbyne complex model system, namely H_5TaCH^{3-}, an octahedral complex related to scheme **17** (2 electrons less) and to compound **18** above, an interaction diagram between TaH_5^{4-} and CH^+ was calculated by the extended Hückel method. The actual purpose of the work was to study the hydrogen transfer reaction shown in **33**. H_5TaCH^{3-} only was discussed as the potential endpoint of this ligand shift, without really considering in detail further aspects of metal carbyne systems.

33

In 1981, three full papers dealing with the electronic structure of transition metal carbyne complexes appeared. The first one came from the Fenske group (38). It was devoted to a computational study of 4 cationic Fischer-type carbyne complexes with $Cr(CO)_5$ and $CpMn(CO)_2$ as the metal fragments and with methylcarbyne, phenylcarbyne, trimethylsilylcarbyne and diethylaminocarbyne as ligands. The computational method used in this study was the so called Fenske-Hall-SCF approximation (FH) (39), a molecular orbital calculation scheme often advertized by its authors as "nonparametrized". It is not an *ab initio* method, but represents an approximation to the Hartree-Fock-Roothaan

LCAO-MO SCF technique, and it has been widely used for transition metal systems since its development. Fixed geometries with reasonable structural estimates, available from X-ray data, were employed in (39). After a discussion of the various carbynes and a detailed review of the familiar MOs of the two metal fragments $Cr(CO)_5$ and $CpMn(CO)_2$, the bonding in $CpMn(CO)_2CMe^+$, $CpMn(CO)_2CSiMe_3^+$, $CpMn(CO)_2CPh^+$ and $(CO)_5CrCNEt_2^+$ is analyzed in detail. The results of the calculations for each of the four carbyne complex cations are portrayed by presenting interaction diagrams between metal fragments and carbyne ligands. Computationally this approach amounts to a transformation of molecular wavefunctions from an atomic orbital basis set to a basis set of fragment orbitals. Within the Fenske-Hall approximation this is somewhat problematic (as in *ab initio*), because in SCF procedures there is no strict transferability of eigenvalues and eigenvectors (MO energies and MOs) between the composite molecule and its constituent fragments, particularly if the latter are charged species. To overcome this difficulty, the so called "frozen orbital approximation" (40) is usually applied in Fenske-Hall work. In this manner the "level scrambling" of fragment MOs, which is due to very different atomic charges in a fragment as such (eg, CR^+), or within the final molecule, does not prevent the possibility of performing a fragment MO interaction analysis.

The bonding in $CpMn(CO)_2CMe^+$, $CpMn(CO)_2CPh^+$ and $CpMn(CO)_2$-$CSiMe_3^+$, as described by Fenske, is basically as represented in scheme **17** above. A strong metal-to-carbyne triple bond is found with practically degenerate π and π^* components, the latter being the LUMO of the complexes. For CPh^+ as the alkylidyne ligand in $CpMn(CO)_2CPh^+$, the phenyl group rotational barrier is therefore virtually non existent. This point is addressed again in a later paper by the same authors (41). For the aminocarbyne complex $(CO)_5CrCNEt_2^+$, the electronic structure of the aminocarbyne fragment, shown in *Figure 1* in section 1 above, does lead to two rather different π-bond components, in accord with the previously described resonance formulation of aminocarbyne ligands as azavinylidenes, ie, **9b** above. Because aminocarbyne complexes were the subject of an independent paper published later in the same year with identical results (18) and are discussed below, we can postpone this point of the Fenske paper. There are several "chemically significant" conclusions in it, however, which need to be mentioned. Within the Fenske-Hall MO model, carbynes are found to be among the best π acceptors, withdrawing electrons better than CO does. Atomic charges, obtained via a Mulliken population analysis, reveal that of all the ligand atoms in the four carbyne complexes, the carbyne carbon is invariably the most negative. This is a significant result, because it had been proposed originally (3c), that the positive charge of cationic carbyne complexes is highly localized at the carbyne carbon atoms. Fenske's MO results show that the qualitative picture of interaction scheme **17**, with the MC bond π^* orbitals as the LUMOs of carbyne complexes, is basically correct. The authors suggest that the orientation of nucleophilic additions to carbyne complexes may be frontier orbital controlled, rather than simply guided by the atomic charge distribution. A net positive charge is suggested simply to enhance the molecule's overall reactivity. No attempts of modelling reactive behavior by calculations were made, however, and although the author's conclusions may be qualitatively quite reasonable and justified, they must await further corroboration either by theory or experiment.

The second paper dealing with the electronic structure of transition metal carbyne complexes was by Hohlneicher et al., who used an extension of the CNDO formalism to study *trans*-Cl(CO)$_4$CrCMe, a representative of the very first "Fischer-type" carbyne complexes. The main purpose of this work (42) was to elucidate the apparent stabilizing effect of the trans halogeno substituent in such neutral d^6-systems. To this end a computational comparison was made with Cr(CO)$_6$ and the hypothetical (CO)$_5$VCMe molecule. Owing to the method employed, fixed model geometries had to be used and again the fragment MO approach helped to explain and understand the results obtained. For (CO)$_5$VCMe the general picture of **17** arises once again. Calculated charges reveal essentialy neutral V, CO and carbyne ligands. Calculated bond orders confirm the metal-to-carbyne multiple bond and exhibit stronger bonds to the cis CO ligands than to the trans carbonyl group.

For *trans*-Cl(CO)$_4$CrCMe the bonding situation is explained by constructing an interaction diagram between Cl(CO)$_4$Cr and CMe radicals. A scheme qualitatively identical to **17** emerges, which only has to be modified insofar as the lonepairs on the Cl ligand now appear as the HOMO of the complex, above the b_2 (xy) level. We come back to *trans*-(halogeno)(CO)$_4$MCR-complexes below. A metal charge of +0.77 is predicted for Cl(CO)$_4$CrCMe from these CNDO calculations – it will be compared to values from other theoretical treatments later. The carbyne carbon from CNDO carries a negative charge of −0.16. According to the final conclusions of this paper, trans-ligands L with a high donor/acceptor ratio, or good π-donors L trans to the carbyne ligands in L(CO)$_4$CrCR complexes, exert a stabilizing effect in such systems. This finding of course reproduces qualitative expectations and experimental results.

The third paper which appeared during 1981 exclusively deals with aminocarbyne complexes of the "Fischer-type", with chromium as the metal (18). A series of 5 chromium carbyne complexes with CNEt$_2^+$ as carbyne ligands had been investigated by X-ray diffraction. Their bonding patterns were analyzed through extended Hückel model calculations in order to clarify the influence of the varying metal fragments in **34–38** upon the bonding parameters to and within the CNEt$_2^+$ ligand.

$Br\text{-}Cr\equiv CNR_2$ **34**

$Ph_3Sn\,Cr\equiv CNR_2$ **35**

$Ph_3P\,Cr\equiv CNR_2$ $\oplus$ **36**

$Br\text{-}Cr\equiv CNR_2$
PPh_3 **37**

$OC\text{-}Cr\equiv CNR_2$ $\oplus$ **38**

$(R = Et, \; \text{—}\bullet = CO)$

Figure 2 gives an EH interaction diagram, using $Cr(CO)_5$ and $C(NH_2)^+$ molecular fragments to construct the MOs of $(CO)_5CrCNH_2^+$ as a model for **38**. This diagram is essentially identical to the one given by Fenske and Kostic in their paper cited above.

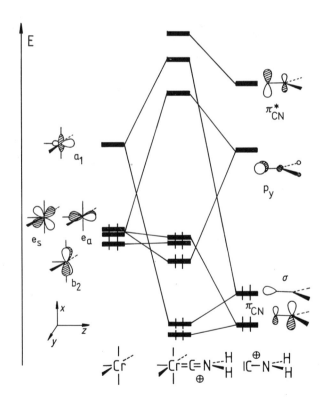

Fig. 2: Extended Hückel interaction diagram of $(CO)_5CrCNH_2^+$.

In contrast to the qualitative scheme **17**, which applies to alkyl- or aryl-substituted carbyne complexes of the L_5MCR-type, only p_y of the aminocarbyne ligand leads to a strong π-bond with the appropriate member of the $Cr(CO)_5$ donor level e_a. Level e_s is only slightly stabilized, owing to the much higher energy and diminished overlap capability of π^*_{CN}, as discussed in the context of **9a/9b** and *Figure 1* (43). The bonding situation in **38** is more of an allene than of an alkyne type, as far as the Cr-C-N chain is concerned. Of the five CO ligands, the overlap population to the one in trans position is weakest, because the metal e_s and (even more so) e_a levels, both of which bond to the trans CO, lose electron density to the p_y (mainly) and π^*_{CN} acceptor levels of the very effective π-acid $C(NH_2)^+$ ($CNEt_2^+$ in **38**). This result explains why cationic alkyl carbyne complexes $(CO)_5CrCR^+$ are too labile to be isolated: for them p_x (viz **5** and **17**), equally potent an acceptor level as p_y, would replace π^*_{CN} of the aminocarbyne and, consequently, the trans CO group would be even better labilized. Model calculations on $(CO)_5CrCMe^+$ confirm this point (18).

Within the series **34–38**, the $(CO)_5Cr$ metal fragment is the weakest π-donor towards the aminocarbyne ligands. It is easy to analyze the effect of going from $(CO)_5Cr$ to, eg, the $Br(CO)_4Cr^-$ unit as the metal moiety. By replacing the axial CO of $(CO)_5Cr$ (C_{4v}) with a Br^- ligand, backbonding in e_s and e_a of **11** is lost and both levels rise in energy. According to the qualitative correlation in **39**, the two e-members of the t_{2g}-based occupied MO triplet of $(CO)_5Cr$ (viz **11**) are further destabilized, because the two p-lonepairs of the halogen mix into these orbitals in an antibonding sense. The b_2 (xy) level remains unaffected by the trans halogen substitution, due to its symmetry.

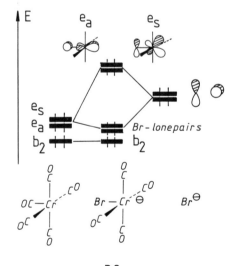

39

Thus a $Br(CO)_4Cr^-$ metal fragment, compared to $(CO)_5Cr$, has a higher lying pair of degenerate donor orbitals (e_s, e_a) with antibonding contributions from the halogen p-lonepairs. Between this MO and b_2 are found two MOs which represent the bonding counterpart of e_s and e_a, the in-phase combination of the halogen lonepairs and of xz and yz at the metal. These two levels, essentially to be viewed as the lonepair orbitals of the halogen atom, still appear as the HOMOs in *trans*-$Br(CO)_4CrCMe$ according to Fenske-Hall MO studies (see below). EH calculations on the model system *trans*-$Br(CO)_4CrCNH_2$ also place them among the highest occupied levels. For $Br(CO)_4Cr^-$ as the metal fragment, the overlap population between Cr and the carbyne carbon atom is greater than for the cationic $(CO)_5CrCNH_2^+$; concomitantly the C-N bond overlap population is reduced since π^*_{CN} receives more electron density from the halogen containing metal fragment. In the language of resonance theory, structure **40b** contributes more in the neutral $Br(CO)_4CrCNR_2$ system, whereas for $(CO)_5CrCNR_2^+$ cations **40a** is more important, as reflected in the experimentally determined Cr-carbyne bond (179.7 pm) of **38**, which is distinctly longer than the one in **34** (172.0 pm).

$$L_n\overset{\ominus}{M}=C=\overset{\oplus}{N}R_2 \longleftrightarrow L_nM\equiv C-\bar{N}R_2$$

$$\textbf{40 a} \qquad\qquad \textbf{40 b}$$

EH calculations were also performed for *trans*-$(PH_3)(CO)_4CrCNH_2^+$ and *mer*-$Br(PH_3)(CO)_3CrCNH_2$ as models of **36** and **37**. For all cases the metal fragment and the amino group compete with each other for π-interaction with the carbyne carbon. In the neutral complex **37** with a trans Br and a cis PR_3 ligand, a preferred rotational orientation of the aminocarbyne (ie of the NR_2 plane) is expected from the model calculations. The Cr-P bond and the NH_2 group (NR_2 group) want to be coplanar, and the electronic origin of this preference, also found nearly precisely in the experimental structure, is analyzed in (18). Overall, the paper gives a consistent description of the bonding in aminocarbyne complexes, which may serve as a basis for predicting the electronic and geometrical structures of related systems.

In 1982, Fenske and Kostic published a second paper on bonding aspects, structures and frontier orbital controlled reactivity for octahedral carbyne complexes (44). Again the Fenske-Hall MO model was employed, and the authors performed model calculations for the series *trans*-$X(CO)_4CrCR$ (X = Cl, Br, I; R = Me, Ph, NEt_2), for *trans*-$PMe_3(CO)_4CrCMe^+$, *mer*-$Br(PMe_3)(CO)_3$-$CrCMe$, various isomers of $Br(CO)_2(CNMe)_2CrCPh$ and $Br(CO)_2[P(OH)_3]_2$-$CrCPh$, as well as for $(PH_3)_2(CO)_2FeCPh^+$. The last was included as a model for an osmium system prepared by Roper. The paper is an extension of the one cited above, applying the same methodology. The bonding description for *trans*-Br(CO)$_4$CrCNEt$_2$ is virtually identical to the one given by Schubert and Hofmann (18) and can be transferred to the Cl and I analogs. For all trans halogeno tetracarbonyl complexes the p-type lonepairs of the halogen ligands are the highest lying MOs. The LUMO in every system is a metal to carbyne π-antibonding level. The trans halogeno tetracarbonyl complexes and their cis-isomers are reported to be of practically identical energy. This has been found computationally at least for the two isomers of I(CO)$_4$CrCMe. Depending on the halogen ligand X, the Cr-X σ-bonding MO is shifted; for X = I it appears as the next level below the HOMO.

The general bonding situation is rather similar for all molecules studied; independent of the specific ligand arrangement it nicely agrees with the qualitative scheme of **17**. A section of the paper is devoted to the reactivity of carbyne complexes towards nucleophiles, and the earlier conclusions with respect to frontier orbitals dominating nucleophilic additions in cationic systems are claimed by the authors to be corroborated for neutral complexes. Several experimental findings are discussed in the light of this interpretation. Although certainly justified from a qualitative point of view, again no unequivocal proof (experimental or theoretical) for frontier orbital governed reactivity can be given. The statement by the authors that the expected primary addition products of nucleophiles to the carbyne carbon may subsequently rearrange, leading to final products not indicative of the true reaction mechanism, is not very helpful in evaluating the significance of their proposed frontier MO control model of reactivity.

A third publication in the series of papers by Kostic and Fenske (45) addresses some conformational problems associated with transition metal carbyne complexes. The rotational barrier about the metal carbyne triple bonds in CpMn-(CO)$_2$CPh$^+$ and BzCr(CO)$_2$CPh$^+$ (Bz = η^6-C$_6$H$_6$) is calculated and is found to be essentially near zero. In other words, there is free rotation of the phenyl rings in both model carbyne complexes, as expected. The main focus of the paper are conformational details of various transition metal carbene complexes, and criteria of maximum overlap and minimum orbital energy are applied to compare different stereoisomers. Conclusions and explanations found in earlier EH work by Hoffmann et al. (21b) on the same theme are augmented and corroborated by the Fenske-Hall MO method.

In the context of the proposed importance of frontier orbitals for controlling the chemical reactivity of carbyne complexes and with respect to the consistently calculated negative charge accumulation on carbyne carbon centers in such molecules, a short paper by Green et al. (46) is of interest. Here is adressed the question of how electrophiles will react with carbyne complexes. EH MO calculations for **41** give net charges on Mo and the carbyne carbon of +0.93 and −0.34, respectively.

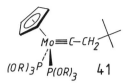

Frontier MO control might thus predict electrophilic attack on the metal (the HOMOs are metal localized MOs), whereas charge control might guide an electrophile to carbon. The experimental results in (46) suggest that a proton initially attacks the carbyne carbon atom under kinetic control. Subsequent reaction steps lead to different thermodynamically controlled end products, one of which is Mo-protonated **41** (a cationic hydrido carbyne complex), the structure of which is reported. So at least here, charge control has been postulated to prevail.

A short theoretical section (without details), dealing with a class of carbyne complexes briefly mentioned in section 3 of this review, appeared in a 1982 paper by Nicholas (47a). A modified version of extended Hückel theory (MEHT) was used to compare relative energies and geometries of the tetrahedral hydroxy-carbyne complex $(CO)_3CoCOH$, and of its isomer $(CO)_3Co(CHO)$ which has a formyl ligand. The tetrahedral geometry (C_{3v}) was found to be 88 kJ/mol more stable than a square planar structure for this d^{10}-ML_3 carbyne complex. This result is in accord with the general bonding description for d^{10}-ML_3CR carbyne complexes mentioned earlier in this account (see section 3). No further details on the electronic structure of the mononuclear hydroxycarbyne complex were given. The author concludes that such species might be possible intermediates in the reactions of metal carbonyl hydrides, particularly in carbon monoxide reductions. Energetically, the formyl isomer (planar) $(CO)_3Co(CHO)$ lies 29 kJ/mol above tetrahedral $(CO)_3CoCOH$, but taking into account the uncertainty of computed total energies under the severe constraints of the study, these numbers have to be viewed with caution.

Two experimental papers, both published in 1983, are relevant to the present discussion. From the generalized bonding scheme of **17**, as well as from all available numerical quantum chemical work, it seems safe to conclude, that "Fischer-type" carbyne complexes typically have a HOMO manifold, which consists essentially of metal-localized (d-orbital) MOs. Moreover, π^*-type LUMOs are centered in the metal-carbyne bond region, mostly localized on carbon. Together this leads to the assignment of the intense low energy transitions found in electronic spectra of such systems (similar to "Fischer-type" carbene complexes (2, 12)) as charge transfer excitations of the metal to ligand type (CTML). This assignment has been verified for two trigonal bipyramidal d^8 osmium phenylcarbyne complexes by Vogler et al. in their study of HCl addition to these molecules (48). The authors suggest that photoexcitation of such carbyne complexes generates a geometrically relaxed CT state with square planar geometry and a bent carbyne ligand, carrying a lonepair at carbon. This intermediate is believed to undergo fast electrophilic attack (by H^+), much faster than the ground state molecule, yielding intermediate carbene complexes by rapidly attaching a chloride to the metal. The close analogy of CR^+ and NO^+ as ligands is again noted in this work, which is nicely consistent with qualitative electronic structure conceptions.

The second piece of experimental work to be mentioned in this context comes from the laboratory of Krüger et al. The electronic deformation density of a prototype "Fischer-carbyne complex", *trans*-$Cl(CO)_4CrCMe$, was experimentally determined by X-ray diffraction. This high resolution X-ray (X-X) study (49) allows an interesting view of the bonding situation of the carbyne ligand. An unusually high electron density within the C-C bond region of the methyl-carbyne unit is reported, while the observed density distribution of the metal-carbyne bond is rather diffuse. The authors interpret these findings in terms of hyperconjugation to the methyl group, which essentially again reflects the requirement of the carbyne carbon p-acceptor orbitals for electron density above and beyond that supplied from the metal fragment. The role played by the nitrogen lonepair in **34–38** and in aminocarbyne complexes in general (viz. **40a/40b**) is assumed, to a much smaller extent, by the CH bonding electrons of the CH_3 group. This hyperconjugative coupling of the Cr-carbon triple bond and the methyl group CH σ-orbitals had also been noted in the CNDO calculations for *trans*-$Cl(CO)_4CrCMe$ (42). A second interesting feature of the X-X study is the observed C_{2v} symmetry of the $Cl(CO)_4Cr$ metal fragment, reflecting the maximum overall symmetry expected for a molecule, in which two C_{3v} and C_{4v} subunits are linked along their two principle axes. Although intermolecular interactions in the crystal are not excluded, the clearly visible C_{4v} to C_{2v} distortion of the metal fragment may indicate appreciable interaction of the methyl group with the metal-carbon triple bond system in accord with the proposed hyperconjugation effect or its equivalent molecular orbital description.

In 1984, a Japanese group published another theoretical treatment of metal-carbon multiply bonded molecules (50), including two carbyne systems, $(CO)_5CrCH^+$ and *trans*-$Cl(CO)_4CrCH$, both with methylidyne as a ligand. *Ab initio* SCF-MOs were calculated by the Hartree-Fock-Roothaan SCF method without geometry optimization (expect for Cr-carbyne distances and the Cr-C-H angle in the cation, for which angles of 180° (better) and 178° were compared). Rather limited basis sets were used throughout, and correlation effects were not taken into account. The Cr-C bond lengths reported from this work (162–165 pm) compare reasonably well with experimental values. The calculated vibrational frequencies are much too low, however. An interaction diagram is derived for each of the two complexes, using a closed shell metal fragment and a CH^+ unit in exactly the way employed several times above. The picture obtained is qualitatively identical to **17** and to the other cases reported in earlier work. In contrast to Fenske-Hall results for *trans*-Cl(CO)$_4$CrCMe (41), the Cr-Cl σ-bonding MO appears next to the Cl lonepair HOMOs. Negative charges for the carbyne carbon atoms in both the neutral and the cationic complex are found, again consistent with earlier results, and the interpretation given is as before. Generally speaking, the net results of this first *ab initio* approach to the electronic structure of "Fischer-type" transition metal carbyne complexes do not extend the previously described outcome of earlier work, as far as chemically significant features are concerned. Owing to the very restricted level of sophistication (on the *ab initio* scale of theory) the calculated numbers have only qualitative meaning anyway. It is clear that Nakatsuji et al. arrive at the same conclusions as previous investigators, and frontier orbital control of reactivity is once again strongly supported by these *ab initio* computations.

While the overall electronic structure description of the closed shell molecules $(CO)_5CrCH^+$ and *trans*-Cl(CO)$_4$CrCH may be quite reasonable on the single determinant level of theory as applied in the work of Nakatsuji et al., all their attempts to compute Cr-CH bond energies corresponding to cleavage of $(CO)_5CrCH^+$ and Cl(CO)$_4$CrCH into fragments of various spin patterns, must be regarded as very unreliable. Correlation effects definitely play a dominant role here and calculations on a level well beyond the Hartree-Fock limit and with larger basis sets are certainly required. The reader is reminded here of what has been said about CH in section 2.

The validity of this caveat is quite apparent in the most recent treatment of the same molecule *trans*-Cl(CO)$_4$CrCH by Poblet et al. (51). These authors performed *ab initio* calculations beyond the Hartree-Fock level, comparing them with single determinant results. The aim of this paper is to discuss changes obtained with respect to the SCF level, using correlated wavefunctions in calculating the metal-carbon triple bond at equilibrium and along the dissociation path. CAS SCF calculations on Cl(CO)$_4$CrCH were performed with a (12, 8, 5/5, 4, 3) basis set for Cr and appropriately balanced large basis sets for Cl, C, O and H, using two different sizes of the CAS SCF space in the correlated calculations. Calculations for the neutral CH and Cl(CO)$_4$Cr fragments in their quartet and doublet states have also been carried out. Referring back to the statements above, it is not surprising that correlation of the six electrons involved in the triple bond lowers the energy, when potential curves for stretching the Cr-C bond are computed. This lowering amounts to around 502 kJ/mol! At the experimental geometry, the triply bonded configuration ($\sigma^2\pi^4$) accounts for only 82 % of the CI expansion within the CAS SCF-6 (six electrons/six MOs) calculation. Going to a CAS SCF-8 expansion, including the bonding and antibonding combinations of the metal xy AO and the π* MOs of the four equatorial COs in the active space, lowers the energy further by 130 kJ/mol. Apparently the agreement between experimental and computed Cr-carbyne bond lengths consistently improves with the quality of the wavefunction. Using a CI wavefunction mixes triple bond π* character into the ground state, thus elongating the Cr-C bond. Even more sensitive to the level of calculation is the triple bond dissociation energy. In contrast to the SCF results of the work by Nakatsuji et al. (50), no positive bond energy at all is found at the SCF level. The two separate fragments $Cl(CO)_4Cr$ and CH, both in their quartet state, are lower in energy by 21 kJ/mol than the carbyne complex at equilibrium. This result, according to Poblet et al., suggests a cancellation of errors introduced by the poor basis set and by neglect of correlation in the earlier Japanese study. The possibility for a dissociation of $Cl(CO)_4CrCH$ into neutral fragments with different spin states was considered by Poblet et al. as well (CH itself correctly comes out as $^2\Pi$, $Cl(CO)_4Cr$ has a $^4A_2$ groundstate). Nevertheless, no inter-fragment net bonding is found at the SCF level. The correlated calculations lead to a dissociation energy of 481 kJ/mol, in good quantitative agreement with the experimental energy of the triple bond in gas phase VCH^+. At the level of theory reached in (51), mutually cancelling errors are still expected by the authors of this study. The basis set superposition error and the neglect of geometrical relaxation should overestimate the bond energy, but the lack of more polarization functions in the basis set and the still incomplete correlation tend to underestimate it.

Partial negative charge of the carbyne carbon, emerging from semiempirical calculations and qualitative bonding concepts, is confirmed both by the French group's SCF and CAS SCF results, but is attributed more to an excess of σ population on the carbyne carbon rather than to strong π-back-donation on the CAS SCF level. The π-bonds according to (51) are essentially nonpolar covalent.

The paper by Poblet et al. clearly emphasizes what has been stated earlier when discussing quantitative electronic structure calculations for the simple CH diatomic: to obtain numerically reliable results, very high levels of quantum methodology must be reached before one can expect to get anything close to a correct description of transition metal carbyne complexes on an absolute basis.

It may seem rather hopeless then to obtain chemically valuable information from theoretical calculations as a tool intended either to guide or interpret experimental work. Fortunately this is not so. Despite difficulties that still hamper or even render impossible numerically accurate work in this area, theoretical chemistry, appropriately employed with the limitations of each methodology in mind, can be of immense help to experimental work. After all, numerically accurate predictions or the precise reproduction of molecular properties do not necessarily imply understanding. Understanding, however, is what is needed in the first place in order to be able to develop new ideas.

This survey of published theoretical work on carbyne complexes concludes by mentioning a recent extended Hückel study by Brower, Templeton and Mingos (52). Octahedral biscarbyne and mixed oxocarbyne complexes of group VI, modelled by *cis*- and *trans*-$W(CH)_2H_4^{4-}$ and *cis*- and *trans*-$W(CH)(O)H_4^{3-}$, were investigated with respect to their valence orbital characteristics. The lack of d-orbitals of appropriate symmetry generates a nonbonding, purely carbyne localized molecular orbital for the *cis*- and *trans*-biscarbyne model $W(CH)_2H_2^{4-}$. This orbital, if filled (empty), causes strong carbanionic (carbocation like) character at the CH groups and therefore infers high kinetic instability for such hypothetical complexes. In oxocarbyne systems $W(CH)(O)H_4^{3-}$, competition occurs between the oxo and methylidyne ligand for π-bonding to the metal. Stabilization is predicted by an umbrella-type distortion of the equatorial ligand set towards the oxygen site in the trans-oxycarbyne complex. Cis-oxocarbyne complexes are even predicted as potentially stable and preparable molecules for a d^4 metal configuration (counting the carbyne as CR^+). The reader may note in passing that this prediction can be also derived from the qualitative interaction diagram **17**: for a d^4 electron count xy is empty and, as already detailed in section 3, a good π-donor (as in "Schrock-systems") dramatically raises this low lying level, thereby stabilizing d^4-carbyne complexes. An oxo ligand (formally O^{2-}) with its filled p levels is ideal for that purpose. Also to be found in (52) are some interesting comments on electron counting conventions in carbyne complexes in general.

5 Outlook

It is pleasing to note that, at least for prototype transition metal carbyne complexes, a consistent qualitative molecular orbital description has built up during the last decade. Clearly more theoretical studies need to be undertaken, in particular for complexes of the "Schrock-type". Both high level *ab initio* work, pushing foreward the frontier of quantitative electronic structure theory, and "applied" quantum chemical studies of the more qualitative, experimentally oriented type are necessary in the future.

The available theoretical work on transition metal carbyne complexes, as collected in this review, provides a good example for a qualitatively consistent, rather method-independent development of an electronic structural concept, which is useful for understanding important aspects of the chemistry in this area.

The most interesting and fascinating questions of the near future will, in my opinion, be related less to "static" electronic or structural problems. Rather, they will arise in the context of understanding or even predicting the chemical reactivity of carbyne complexes. In fact such studies have already begun to appear – a detailed MO analysis of coupling two carbyne ligands on a mononuclear or a binuclear transition metal fragment as a template (53), published by R. Hoffmann and coworkers, is an early and leading example. Experimental cases of carbyne-carbonyl coupling to ketenyl ligands (54) and, more recently, of carbyne-carbyne coupling to form metal bound alkynes (6b) have been found and studied. The qualitative "prescriptions" from Hoffmann's work, relating the d-electron count to the "allowed" or "forbidden" nature of carbyne coupling steps, seem to hold well in many cases, although all conclusions were derived from the "naked" metal biscarbyne model $W(CH)_2$. We have ourselves studied more realistic cases recently, like carbyne-CO coupling in **42** or carbyne- carbyne coupling in **43** and **44**, calculating total energy surfaces and studying the effect of incoming ligands upon the coupling step of such C_1-fragments (55). A detailed discussion would exceed the scope of this account and will be given elsewhere.

Incidentally, there is an extremely fascinating and very recent piece of experimental work, which indicates some general importance of carbyne complexes for coupling reactions of CO ligands bound to transition metal centers. Lippard and coworkers were able to uncover the entire reaction mechanism of their exciting reductive coupling of two carbon monoxides within the coordination sphere of a seven coordinate tantalum complex, which had been reported earlier to yield metall coordinated bis-siloxyalkyne ligands (56). All intermediates were isolated and structurally characterized. The key compound has turned out to be an octahedral d^6-tantalum complex **45** with a siloxycarbyne and a carbonyl ligand cis to each other (57). These two fragments subsequently undergoe carbon-carbon bond formation (possibly to an intermediate η^2-ketenyl ligand (58)), finally leading to a coordinated acetylene. These findings perhaps indicate a general opportunity for utilizing coordinated CO ligands synthetically by transforming them into oxycarbyne groups.

Another very important area, which certainly would benefit from theoretical studies, is the field of alkyne metathesis. It is somewhat surprising that, to the best of our knowledge, no theoretical investigation exists, which tackles the

45

reactivity of carbyne complexes towards alkynes, although a very recent GVB study by Goddard et al. (26) has dealt with $L_nMC_3H_3$ isomers.

The increasing importance of tailor-made transition metal organometallics as stochiometric or even catalytic reagents in organic synthesis no doubt will also offer opportunities for an application of carbyne complexes – provided we can gain a sufficient amount of understanding of their structures, electronic properties and reactive behavior by engaging ourselves in the fascinating interplay between theory and experiment.

6 Acknowledgement

I am indebted to the members of my research group, and to Dr. A. C. Filippou and Dr. J. Okuda for discussions of various aspects of this account. I am especially grateful to Prof. S. J. Lippard for reading and polishing this manuscript, for his comments and for disclosing experimental work prior to publication. The skillfull typing of the manuscript by Mrs. I. Gruber is gratefully acknowledged.

7 References

(1) E. O. Fischer, G. Kreis, C. G. Kreiter, J. Müller, G. Huttner, H. Lorenz, *Angew. Chem.* **85** (1973) 618; *Angew. Chem. Int. Ed. Engl.* **12** (1973) 564.

(2) For the chemistry of transition metal carbene complexes see: K. H. Dötz, H. Fischer, P. Hofmann, F. R. Kreissl, U. Schubert, K. Weiss, *Transition Metal Carbene Complexes,* Weinheim: Verlag Chemie, 1983, and references therein.

(3) E. O. Fischer, U. Schubert, *J. Organomet. Chem.* **100** (1975) 59; b) E. O. Fischer, *Adv. Organomet. Chem.* **14** (1976) 1; c) E. O. Fischer, U. Schubert, H. Fischer, *Pure Appl. Chem.* **50** (1978) 857; d) M. A. Gallop, W. R. Roper, *Adv. Organomet. Chem.* **25** (1986) 121; e) R. R. Schrock, *Acc. Chem. Res.* **19** (1986) 342.

(4) H. P. Kim, R. J. Angelici, *Adv. Organomet. Chem.* **27** (1987) 51.

(5) J. H. Wengrovius, J. Sancho, R. R. Schrock, *J. Am. Chem. Soc.* **103** (1981) 3932.

(6) For most recent examples see: a) G. A. McDermott, A. Mayr, *J. Am. Chem. Soc.* **109** (1987) 580; b) A. Mayr, G. A. McDermott, A. M. Dorries, D. van Engen, *Organometallics* **6** (1987) 1503, and references to earlier work therein.

(7) D. Seyferth, *Adv. Organomet. Chem.* **14** (1976) 97.

(8) F. G. A. Stone, *Angew. Chem.* **96** (1984) 85; *Angew. Chem. Int. Ed. Engl.* **23** (1984) 89.

(9) R. Hoffmann, *Angew. Chem.* **94** (1982) 725; *Angew. Chem. Int. Ed. Engl.* **21** (1982) 711.

(10) E. O. Fischer, N. Q. Dao, W. R. Wagner, *Angew. Chem.* **90** (1978) 51; *Angew. Chem. Int. Ed. Engl.* **17** (1978) 50.

(11) T. A. Albright, J. A. Burdett, M. H. Whangbo, *Orbital Interactions in Chemistry,* New York: Wiley, 1985.

(12) P. Hofmann, *Electronic Structures of Transition Metal Carbene Complexes,* page 113 ff in ref. (2) and literature cited therein.

(13) G. Herzberg, J. W. Johns, *Astrophys.* **158** (1969) 399.

(14) W. J. Hehre, L. Radom, P. v. R. Schleyer, J. A. Pople, *Ab Initio Molecular Orbital Theory,* New York: Wiley, 1986.

(15) G. C. Lie, J. Hinze, B. Liu, *Chem. Phys.* **57** (1972) 625; **59** (1973) 1872.

(16) P. R. Sharp, S. J. Holmes, R. R. Schrock, M. R. Churchill, H. J. Wasserman, *J. Am. Chem. Soc.* **103** (1981) 965.

(17) The slight destabilization of p_y of $C(NH_2)^+$ compared to p_y and p_x of $C(CH_3)^+$ is a consequence of hyperconjugation with the NH_2 group (antibonding interaction with an NH_2-orbital). This seems somewhat unusual

because, for electronegativity reasons, NH bonds should be less efficiently hyperconjugating than CH bonds of CH_3. The result simply arises from the different C-C vs. C-N distances and CCH vs. CNH angles chosen in the EH calculations of $C(CH_3)^+$ and $C(NH_2)^+$, which were taken from available experimental structures of appropriate carbyne complexes. See also ref. (18).

(18) U. Schubert, D. Neugebauer, P. Hofmann, B. E. R. Schilling, H. Fischer, A. Motsch, *Chem. Ber.* **114** (1981) 3365.

(19) Note that p_y is more or less exclusively localized at the carbyne C, whereas π^*_{CN} is delocalized. This difference leads to better overlap capability of p_y.

(20) For the first anionic carbyne complexes see for example: E. O. Fischer, A. C. Filippou, H. G. Alt, U. Thewalt, *Angew. Chem.* **97** (1985) 215; *Angew. Chem. Int. Ed. Engl.* **24** (1985) 203.

(21) The electronic structure of a d^6-CpML$_2$ metal fragment has been described in detail in: a) P. Hofmann, *Angew. Chem.* **89** (1977) 551; *Angew. Chem. Int. Ed. Engl.* **16** (1977) 536; b) B. E. R. Schilling, R. Hoffmann, D. L. Lichtenberger, *J. Am. Chem. Soc.* **101** (1979) 585.

(22) S. M. Rocklage, R. R. Schrock, M. R. Churchill, H. J. Wasserman, *Organometallics* **1** (1982) 1332.

(23) a) M. R. Churchill, Y.-J. Li, L. Blum, R. R. Schrock, *Organometallics* **3** (1984) 109; b) M. R. Churchill, Y.-J. Li, *J. Organomet. Chem.* **282** (1985) 239.

(24) a) A. R. Rossi, R. Hoffmann, *Inorg. Chem.* **14** (1975) 365. b) Another example of a trigonal bipyramidal metal fragment, carrying an aminocarbyne ligand, is $PPh_3(CO)_3FeCNPr^i_2{}^+$: E. O. Fischer, J. Schneider, D. Neugebauer, *Angew. Chem.* **96** (1984) 814; *Angew. Chem. Int. Ed. Engl.* **23** (1984) 820.

(25) F. A. Cotton, W. Schwotzer, E. S. Shamshoum, *Organometallics* **3** (1984) 1770.

(26) E. V. Anslyn, M. J. Brusich, W. A. Goddard III, *Organometallics* **7** (1988) 98.

(27) T. A. Albright, P. Hofmann, R. Hoffmann, *J. Am. Chem. Soc.* **99** (1977) 7546; See also ref. (11) and (26) and related citations therein.

(28) It should be mentioned, that our knowledge of ML$_3$ molecular fragments is closely related to the chemistry of M$_2$L$_6$ dimers with M-M triple bonds. A considerable amount of theoretical work has been published in this context, see: F. A. Cotton, R. A. Walton, *Multiple Bonds Between Metal Atoms,* New York: Wiley, 1982. Another description of ML$_3$ fragments is available in: T. Ziegler, A. Rauk, *Inorg. Chem.* **18** (1979) 1755.

(29) As in **11**, **17** and **20**, no specific electronic state is indicated by **24**. Only the number of electrons present is specified. Occupation of levels and the spin

arrangement of a true d^4-$ML_3(C_{3v})$ are irrelevant, as we use ML_3 only as a building block with a deliberately fixed geometry, appropriate for the construction of the final molecule.

(30) P. Hofmann, unpublished model calculations for $(HO)_3WCH$.

(31) a) M. H. Chisholm, D. M. Hoffman, J. C. Huffman, *Inorg. Chem.* **22** (1983) 2903; b) For very related V-N analogs see: D. D. Devore, J. D. Lichtenhan, R. Takusagawa, E. A. Maatta, *J. Am. Chem. Soc.* **109** (1987) 7408 and reference therein.

(32) J. H. Osborne, A. L. Rheingold, W. C. Trogler, *J. Am. Chem. Soc.* **107** (1985) 7945.

(33) Obviously **30** carries one electron more than **31**. This unpaired additional electron has been shown to be essentially metal localized (in $1a_1$) by ESR.

(34) J. W. Lauher, R. Hoffmann, *J. Am. Chem. Soc.* **98** (1976) 1729; see also references therein and in (11).

(35) If in **32**, the π-bonding interaction of p_x with xz of Cp_2M would be so strong as to shift this second π-bond MO (LUMO in **32**) below the $1a_1$ (y^2) HOMO, an electronically unstable situation with a very small HOMO-LUMO gap would still result. Thus the conclusion of an unfavorable bonding relationship is independent of the precise location of the LUMO in **32**.

(36) E. O. Fischer, A. C. Filippou, private communication.

(37) R. J. Goddard, R. Hoffmann, E. D. Jemmis, *J. Am. Chem. Soc.* **102** (1980) 7667.

(38) N. M. Kostic, R. F. Fenske, *J. Am. Chem. Soc.* **103** (1981) 4677.

(39) M. B. Hall, R. F. Fenske, *Inorg. Chem.* **11** (1972) 768.

(40) D. L. Lichtenberger, R. F. Fenske, *J. Chem. Phys.* **211** (1976) 995.

(41) N. M. Kostic, R. F. Fenske, *J. Am. Chem. Soc.* **104** (1982) 3879.

(42) D. Saddei, H.-J. Freund, G. Hohlneicher, *J. Organomet. Chem.* **216** (1981) 235.

(43) In addition, π_{CN} mixes into this level in an antibonding way from below, keeping it from being further stabilized.

(44) N. M. Kostic, R. F. Fenske, *Organometallics* **1** (1982) 489.

(45) N. M. Kostic, R. F. Fenske, *J. Am. Chem. Soc.* **104** (1982) 3879.

(46) M. Green, A. G. Orpen, I. D. Williams, *J. Chem. Soc. Chem. Commun.* (1982) 493.

(47) a) K. M. Nicholas, *Organometallics* **1** (1982) 1713;
b) E. O. Fischer, J. R. Schneider, *J. Organomet. Chem.* **295** (1985) C 29.

(48) A. Vogler, J. Kisslinger, W. J. Roper, *Z. Naturforsch.* **38b** (1983) 1506.

(49) a) C. Krüger, R. Goddard, K. H. Claus, *Z. Naturforsch.* **38b** (1983) 1431;
b) R. Goddard, C. Krüger, in: *Electron Distributions and The Chemical Bond:* P. Coppens, M. Hall, (eds.) New York: Plenum Press, 1982.

(50) J. Ushio, H. Nakatsuji, T. Yonezawa, *J. Am. Chem. Soc.* **106** (1984) 5892.

(51) J. M. Poblet, A. Strich, R. Wiest, M. Benard, *Chem. Phys. Lett.* **126** (1986) 169.

(52) D. C. Brower, J. L. Templeton, D. M. P. Mingos, *J. Am. Chem. Soc.* **109** (1987) 5203.

(53) a) R. Hoffmann, C. N. Wilker, O. Eisenstein, *J. Am. Chem. Soc.* **104** (1982) 632; b) C. N. Wilker, R. Hoffmann, O. Eisenstein, *Nouv. J. Chim.* **7** (1983) 535.

(54) a) K. R. Birdwhistell, T. L. Tonker, J. L. Templeton, *J. Am. Chem. Soc.* **107** (1985) 4474; See also the references therein and in (6b) and (53). b) For coupling of carbyne units on multimetal templates, ie, surfaces, see: C. Zheng, Y. Apeloig, R. Hoffmann, *J. Am. Chem. Soc.* **110** (1988) 749.

(55) P. Hofmann, H. R. Schmidt, M. Frede, unpublished results.

(56) P. A. Bianconi, R. N. Vrtis, Ch. Pulla Rao, I. D. Williams, M. P. Engeler, S. J. Lippard, *Organometallics* **6** (1987) 1968.

(57) R. N. Vrtis, Ch. Pulla Rao, S. W. Warner, S. J. Lippard, *J. Am. Chem. Soc.* **110** (1988) 2669.

(58) The first observation of carbyne/CO coupling yielding coordinated η^2-ketenyl ligands (subsequently found by other workers to occur in various related systems, *viz.* ref. (6) and (54) above) was reported more than a decade ago: F. R. Kreissl, A. Frank, U. Schubert, T. L. Lindner, G. Huttner, *Angew. Chem.* **88** (1976) 649; *Angew. Int. Ed. Engl.* **15** (1976) 632; F. R. Kreissl, P. Friedrich, G. Huttner, *Angew. Chem.* **89** (1977) 110; *Angew. Chem. Int. Ed. Engl.* **16** (1977) 102; F. R. Kreissl, K. Ebert, W. Uedelhoven, *Chem. Ber.* **110** (1977) 3782.

Selected Reactions
of Carbyne Complexes

Fritz R. Kreissl

1 Introduction

In 1973, E. O. Fischer et al. (1) reported the first successful synthesis of a transition metal carbyne complex by the reaction of pentacarbonyl(methoxyphenylcarbene)chromium with boron trifluoride. Subsequently, many other carbyne complexes have been synthesized by the classical route of E. O. Fischer or *via* new preparative methods (2–4). The exploration of the chemistry of these carbyne complexes has attracted many active research groups. Moreover, transition metal carbyne complexes have become even more important as starting materials for the synthesis of new organometallic complexes.

This article will be concerned with selected examples of such applications, and will mainly concentrate on reactions of carbyne complexes of the Fischer type (A), which are characterized by an electrophilic carbyne carbon. In this respect, their reaction behavior is in contrast to the Schrock alkylidyne (5,6) complexes (C). The dicarbonyl(cyclopentadienyl)carbyne complexes of molybdenum and tungsten (B) exhibit nucleophilic properties of the metal carbon triple bond and can be placed between the Fischer and the Schrock types:

$$Hal(CO)_4M \equiv C–R \qquad C_5H_5(CO)_2M \equiv C–R \qquad Hal(PMe_3)_4M \equiv C–R$$
$$(A) \qquad\qquad (B) \qquad\qquad (C)$$

The chemical properties of these carbyne complexes are indicated in Scheme 1.

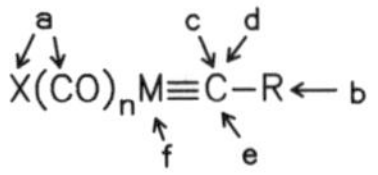

$$X(CO)_nM \equiv C–R$$

Scheme 1.

a) Substitution, b) modification of the carbyne side chain, c) addition of a nucleophile, d) addition of an electrophile, e) transfer of the carbyne ligand, f) oxidation or reduction of the metal.

2 Substitution Reactions

2.1 Substitution of Non-Carbonyl Ligands

Substitution reactions of non-carbonyl ligands *trans* or *cis* to the carbyne unit occur readily with most carbyne complexes to yield a variety of new carbyne complexes. One of the best studied substitution reactions for the *trans*-ligand is the displacement of a halide ligand by free halide ions or dichloromethane (7–11)].

$$X(CO)_2LL'M \equiv CR + Y^- \rightarrow Y(CO)_2LL'M \equiv CR + X^-$$

M = Cr, Mo, W; X = Cl, Br, $AsPh_2$; Y = Br, I, CN, $AsPh_2$; LL' = bipy, phen; R = Ph, Tol, C_5H_7, Fc, $SiPh_3$, NEt_2

For $X(CO)_4 W \equiv CR$ the reversible halide ligand exchange could be observed (11, 12). These substitution reactions follow a second-order rate law.

$$X(CO)_4W \equiv CR + Y^- \rightleftharpoons Y(CO)_4W \equiv CR + X^-$$

R = Ph, Tol, $4\text{-}C_6H_4CF_3$; X, Y = Cl, Br, I

The formal exchange of the bromide by a cyanide group in *trans*-$Br(CO)_2LL'W \equiv CNEt_2$ (LL' = bipy) leads to a replacement of the bromide ligand by the chelate ligand. The CN group occupies a *cis* coordination site relative to the carbyne ligand (13); in the case of a bromide – diphenyl arsenide exchange, no rearrangement is observed (14).

On the other hand, with anions like $[(CO)_5MEPh_2]^-$ (M = Cr, Mo, W; E = P, As, Sb) the substitution of the bromide occurs to give neutral diethylaminocarbyne complexes with a bridging element of the 15th group (15). These complexes exist as a mixture of *cis* and *trans* isomers.

$$Br(CO)_2LL'W \equiv CNEt_2 + [(CO)_5MEPh_2]^- \rightarrow$$
$$(CO)_5M\text{-}EPh_2\text{-}(CO)_2LL'W \equiv CNEt_2 + Br^-$$

M = Cr, Mo, W; LL' = bipy, phen;
E = P, As, Sb

These substitution reactions can be extended to other metal carbyne complexes with ligands like benzoylpentacarbonylmetallato (M = Cr, W) (16, 17) or tetrafluoroborato. The BF_4-group is coordinated to the metal and can easily be replaced by water (18), phosphines, cyanide, isothiocyanide (19), arsines or iso-nitrile (20) to yield neutral

$$(BF_4)(CO)_4W \equiv CR + Nu \rightarrow Nu(CO)_4W \equiv CR + \dots$$

Nu = SCN⁻, CN⁻; R = Me, Ph

or cationic carbyne complexes.

$$(BF_4)(CO)_4W \equiv CPh + Nu \rightarrow [Nu(CO)_4W \equiv CPh][BF_4]$$

Nu = PPh₃, AsPh₃, CNBuᵗ

$$mer\text{-}(BF_4)(CO)_3(PMe_3)Cr \equiv CMe + Nu \rightarrow$$
$$mer\text{-}[Nu(CO)_3(PMe_3)Cr \equiv CMe][BF_4] + \dots$$

Nu = H₂O, PPh₃

The conversion of the neutral carbonyl(chloro)(tolylcarbyne)bis(triphenyl-phosphine)osmium into cationic carbyne complexes can easily be achieved by substitution of the chloro ligand with carbon monoxide or isonitrile, respectively (21).

$$Cl(CO)(PPh_3)_2Os \equiv CTol + L + AgClO_4 \rightarrow$$
$$[CO(PPh_3)_2LOs \equiv CTol]ClO_4 + AgCl$$

L = CO, CNR

A special type of "substitution" of the *trans* ligand can be seen in the reaction of *trans*-halo-(tetracarbonyl)carbyne complexes of chromium and tungsten with dichloromethane. A methylenechloride-stabilized $Cr(CO)_5$ moiety, synthesized by thermal decomposition of *trans*-bromo(tetracarbonyl)cyclopropylcarbyne-chromium (22) or irradiation of chromiumhexacarbonyl (23) adds to the halo ligand of the carbyne complex to form new halogen-bridged dinuclear carbyne complexes. They are readily cleaved, even by weak donors, such as ether, tetra-hydrofuran, triphenylphosphine, or bromide to form the corresponding carbyne complex and the pentacarbonyl derivative of the donor.

$$X(CO)_4M \equiv CR + \text{``}Cr(CO)_5\text{''} \rightarrow (CO)_5Cr\text{-}X\text{-}(CO)_4M \equiv CR$$

M = Cr, W; R = Me, Ph, C₃H₅; X = Cl, Br, I

The halide in $X(CO)_4M \equiv CR$ can also be substituted by sodium pentacarbo-nylmanganate and -rhenate (10, 24–26) or tetracarbonylcobaltate (27) yielding carbyne complexes with metal-metal bonds.

$$X(CO)_4M \equiv CR + NaM'(CO)_{4/5} \rightarrow (CO)_{4/5}M'\text{-}M(CO)_4 \equiv CR + ..$$

M = Cr, Mo, W; X = Cl, Br; R = Ph, SiPh$_3$, menthyl; M' = Mn, Re, Co

Analogously, sodium tricarbonyl(cyclopentadienyl)molybdate or -tung-stenate react with *trans*-bromo(tetracarbonyl)phenylcarbynetungsten to yield binuclear carbyne complexes (25).

$$Br(CO)_4W \equiv CPh + NaM'(CO)_3Cp \rightarrow$$
$$Cp(CO)_3M'\text{-}W(CO)_4 \equiv CPh + NaBr$$

M' = Mo, W

2.2 Combined Halogen and Carbonyl Ligand Substitution

trans-Halo(tetracarbonyl)carbyne complexes of chromium (28), molybde-num and tungsten $X(CO)_4M \equiv CR$(Mo: R = Me, Tol (29); W: Me, C$_5$H$_7$ (30), C$_3$H$_5$ (31), Ph, Tol, C$_6$H$_4$OMe, Mes, C$_6$H$_3$-3-Br-4-OMe, Fc, NEt$_2$ (32,33), SiPh$_3$ (10)) react with sodium cyclopentadienyl, potassium pentamethylcyclopenta-dienyl (34), lithium indenyl or hydrotris (1-pyrazolyl)borate (= HB(pz)$_3$) (35, 36) with concomitant replacement of the halide and two carbonyl ligands to form new dicarbonyl-substituted carbyne complexes.

$$X(CO)_4M \equiv CR + M'Ar \rightarrow Ar(CO)_2M \equiv CR + M'X + 2 CO$$

M = Cr, Mo, W; X = Cl, Br; R = Me, Ph, Tol, Mes, C$_6$H$_4$OMe, C$_6$H$_3$-3-Br-4-OMe, Fc, NEt$_2$, SiPh$_3$; M' = Li, Na; Ar = Cp, Cp*, Ind, HB(pz)$_3$.

Formal replacement of the cyclopentadienyl ligand in $Cp(CO)_2W \equiv CR$ by electronically equivalent carborane groups, such as 1,2-C$_2$B$_9$H$_9$Me$_2$ (37), yields anionic carbyne complexes with comparable properties.

$$Br(CO)_4W \equiv CTol + Na_2[7,8\text{-}C_2B_9H_9Me_2] \rightarrow$$
$$[(1,2\text{-}C_2B_9H_9Me_2)(CO)_2W \equiv CTol]^- + ...$$

In a similar way *trans*-bromo(tetracarbonyl)organylcarbyne complexes of molybdenum and tungsten react with the sodium salts Na $[CpCo(P(O)R'_2)_3]$ (R' = OMe, OPr^i) to give neutral carbyne complexes with the composition $[CpCo-(P(O)R_2)_3](CO)_2M \equiv CR$. The anions $[CpCo(P(O)R'_2)_3]^-$ act as tris-chelating oxygen ligands in this reaction (38).

$$Br(CO)_4M \equiv CR + Na[CpCo(P(O)R'_2)_3] \rightarrow$$

$$[CpCo(P(O)R'_2)_3](CO)_2M \equiv CR + 2\ CO + NaBr$$

M = Mo, W; R = Ph, Tol; R' = OMe, OPr^i

When *trans*-$X(CO)_2LL'W \equiv CNEt_2$ (X = Br, I; LL' = bipy, phen) reacts with *cis*$[Mo(CO)_4(PPh_2K)_2]$, an anionic carbyne complex with a PP-chelate ligand (39) is formed, which reacts with carbon dioxide with CO_2 insertion into the tungsten-carbon triple bond.

$$X(CO)_2LL'W \equiv CNEt_2 + Mo(CO)_4 (PPh_2K)_2 \rightarrow$$

2.3 Carbonyl Substitution and Rearrangement Reactions

Reactions of *trans*-halo(tetracarbonyl)carbyne complexes of chromium, molybdenum and tungsten with different donor molecules result in the replacement of one or more carbonyl ligands yielding moderately stable carbyne complexes (40), which at the same time contain coordinatively labile ligands of high reactivity. Such complexes are also available from the reaction of *cis*-trimethylphosphine-substituted carbene complexes with boron trihalides (41).

$$Br(CO)_4M \equiv CPh + L \rightarrow Br(CO)_3LM \equiv CPh + CO$$

M = Cr, W; L = P(Pr^i)$_3$, PPh$_3$, AsPh$_3$, SbPh$_3$, P(OPh)$_3$

Reaction with an excess of the donor, or the use of bidentate ligands results in the substitution of two carbonyl groups (40, 42).

$$Br(CO)_4M \equiv CR + LL' \rightarrow Br(CO)_2LL'M \equiv CR + 2\ CO$$

M = Cr, W; R = Me, Ph; LL' = 2 py, 2 PPh$_3$, 2 AsPh$_3$,
2 P(OPh)$_3$, 2 CNBu^t, phen, bipy

According to IR investigations the remaining CO ligands occupy *cis* positions. One exception is the reaction product from Br(CO)$_4$Cr $\equiv$ CPh and P(OPh)$_3$ in which there is *trans* coordination of the two carbonyl groups (40). With trimethylphosphine, the main reaction is addition of the phosphine to the carbyne carbon to give phosphinocarbene complexes (43) (eg, see Chapter 4.1).

A very interesting synthetic route to substituted carbyne complexes starting with anionic acyl pentacarbonylmetal complexes, [M(C(O)R)(CO)$_5$]$^-$, and an equivalent amount of phosgene or oxalyl halide was demonstrated by A. Mayr (44, 45). The solutions of the resulting tetracarbonylcarbyne complexes can be used directly for further reactions. Addition of excess nitrogen-donor ligands leads to formation of bis-substituted carbyne complexes. The pyridine-substituted carbyne complexes easily undergo

$$X(CO)_4M \equiv CR + LL' \rightarrow X(CO)_2LL'M \equiv CR + 2\ CO$$

M = Cr, Mo, W; X = Cl, Br; R = Me, Ph; L = 2 py, bipy, tmeda

further displacement of the pyridine ligands. With bis(diphenylphosphino)-ethane (dppe) the corresponding Cl(CO)$_2$(dppe)W $\equiv$ CR (R = Me, Ph) complexes are formed (45). Trimethylphosphine-substituted carbyne complexes can easily be obtained by reacting the pyridine complex with trimethylphosphine (45).

$$Cl(CO)_2(py)_2W \equiv CPh + 2\ PMe_3 \rightarrow Cl(CO)_2(PMe_3)_2W \equiv CPh + 2\ py$$

With trimethylphosphite the formal substitution of three or four ligands can be achieved when starting from tetracarbonyl or dipyridine-substituted carbyne complexes (46).

$$X(CO)_2L_2M \equiv CPh + 3\ P(OMe)_3 \rightarrow X(CO)[P(OMe)_3]_3M \equiv CPh + 2\ L + CO$$

M = Cr, Mo, W; X = Cl, Br; L = CO, py

In neat trimethylphosphite, substitution of the remaining carbonyl ligand by trimethylphosphite finally proved successful for the molybdenum and tungsten compounds.

$$Cl(CO)[P(OMe)_3]_3M \equiv CPh + P(OMe)_3 \rightarrow Cl[P(OMe)_3]_4M \equiv CPh + CO$$

Reaction of the cationic carbyne complex [*mer*-(dppe)(CO)$_3$W $\equiv$ CCH$_2$Ph]-[BF$_4$] with tetraethylammonium halides leads to the substitution of one carbon monoxide ligand by halide, and isomerisation to form neutral Fischer-type *trans*-halocarbyne complexes (47).

$$[mer\text{-}(dppe)(CO)_3W \equiv CCH_2Ph][BF_4] + [NEt_4]X \rightarrow$$

$$trans\text{-}X(dppe)(CO)_2W \equiv CCH_2Ph + [NEt_4][BF_4] + CO$$

X = Cl, Br, I

Heating a solution of [*mer*-(dppe)(CO)$_3$W $\equiv$ CCH$_2$Ph][BF$_4$] in the absence of the tetraethylammonium halides causes simple loss of carbon monoxide and iso-merisation to form [(dppe)(CO)$_2$W $\equiv$ CCH$_2$Ph][BF$_4$], with both dppe donor atoms and the two carbonyl ligands *cis* to the carbyne ligand (47).

$$[mer\text{-}(dppe)(CO)_3W \equiv CCH_2Ph][BF_4] \rightarrow$$

$$[(dppe)(CO)_2W \equiv CCH_2Ph][BF_4] + CO$$

Subsequent addition of neutral ligand molecules like trimethylphosphine, water or acetone leads readily to cationic carbyne complexes,

$$[(dppe)(CO)_2W \equiv CCH_2Ph][BF_4] + L \rightarrow$$

$$[trans\text{-}(dppe)(CO)_2LW \equiv CCH_2Ph][BF_4]$$

L = PMe$_3$, Me$_2$CO, H$_2$O

whereas with tetraethylammonium halides, neutral *trans*-halocarbyne complexes can be obtained (47).

$$[(dppe)(CO)_2W \equiv CCH_2Ph]^+ + X^- \rightarrow trans\text{-}X(dppe)(CO)_2W \equiv CCH_2Ph$$

X = F, Cl, Br, I

The bispyridine-substituted carbyne complexes can successfully be used for the synthesis of anionic carbyne complexes (48). On reaction with tetraethylammonium chloride in dichloromethane, the anionic carbyne derivatives are formed.

$$Cl(CO)_2(py)_2W \equiv CR + [NEt_4]Cl \rightarrow [NEt_4][Cl_2(CO)_2(py)W \equiv CR] + py$$

R = Me, Ph

These derivatives undergo carbonyl substitution in the presence of the alkenes fumaronitrile or maleic anhydride.

$$[Cl_2(CO)_2(py)W \equiv CR]^- + >C=C< \rightarrow$$
$$[Cl_2(CO)(>C=C<)(py)W \equiv CR]^- + CO$$

R = Me, Ph; >C=C< = fumaronitrile, maleic anhydride.

Another synthetic approach to anionic carbyne complexes was described by E. O. Fischer. The reaction of *trans*-halo(tetracarbonyl)carbyne complexes of tungsten with chlorides or bromides leads, with the displacement of one carbonyl ligand, to new anionic dihalo(tricarbonyl)carbyne complexes (49).

$$X(CO)_4W \equiv CNR_2 + [NEt_4]X \rightarrow [NEt_4][X_2(CO)_3W \equiv CNR_2] + CO$$

X = Cl, Br; R = c-C_6H_{11}, Pr^i

Treatment of *trans*-I(CO)$_2$LL'W$\equiv$CNEt$_2$ (LL' = bipy, phen) with Na$_2$[NC(S)C=C(S)CN] causes displacement of the iodide ligand and elimination of the chelating ligand LL', to give a five-coordinated anionic carbyne complex with a chelating S-C(CN)=C(CN)-S ligand (50).

The first dianionic transition metal carbyne complex was obtained by the reaction of *trans*-(TolS)(CO)$_4$W $\equiv$ CNEt$_2$ with [NEt$_4$]SCN by replacement of the *trans*-TolS and two carbonyl ligands (51).

$$TolS(CO)_4W \equiv CNEt_2 + 3\ SCN^- \rightarrow [(SCN)_3(CO)_2W \equiv CNEt_2]^{2-} + \ldots$$

A very easy, but nevertheless important reaction, is the substitution of the *trans*-carbonyl ligand in cationic aminocarbyne complexes to yield a wide variety of cationic or neutral carbyne compounds (52).

$$[(CO)_5Cr \equiv CNEt_2]^+ + PPh_3 \rightarrow [PPh_3(CO)_4Cr \equiv CNEt_2]^+ + CO$$

$$[(CO)_5M \equiv CNR_2]^+ + X^- \rightarrow X(CO)_4M \equiv CNR_2 + CO$$

M = Cr (53, 54), Mo (55), W (55–58); R = Me,
Et, ½ C$_5$H$_{10}$, Ph; X = Cl, Br, I, BF$_4$, SbF$_6$, SbCl$_6$, CN,
OCN, SCN, SeCN, OPh, SPh, STol, SePh, SeC$_6$H$_4$F, SeC$_6$H$_4$CF$_3$.

The reaction of [(CO)$_5$Mo ≡ CNEt$_2$]$^-$ salts, at low temperature, with bromide, iodide or thiocyanate, e. g., yields the neutral *trans*-substituted carbyne complexes. On the contrary, cyanate or trifluoromethylphenylselenolate form binuclear biscarbyne complexes (55). Analogous complexes can be obtained from *trans*-X(CO)$_4$M ≡ CNR$_2$ on warming a solution of the compound

2 X(CO)$_4$M ≡ CNR$_2$ → (μ-X)$_2$[(CO)$_3$M ≡ CNR$_2$]$_2$ + 2 CO

M = Cr, Mo, W; R = Me, Et; X = Br, I, SCN, OCN, SPh, SePh, SeC$_6$H$_4$F, SeC$_6$H$_4$CF$_3$.

or by reacting pentacarbonyl(diethylamino)carbynetungsten tetrafluoroborate with azide (55).

2 [(CO)$_5$W≡CNEt$_2$][BF$_4$] + 2 N$_3^-$ ⟶

$$Et_2NC≡(CO)_3W \overset{N_3}{\underset{N_3}{\diagup\diagdown}} W(CO)_3≡CNEt_2 + ...$$

Other halide-bridged binuclear complexes are derived from the reactions of *trans*-halo(tetracarbonyl)carbyne complexes of tungsten with methylenebis(diphenylarsine) (59, 60) or methylaminobis(difluorophosphine) (61). In the first reaction, with methylenebis(diphenylarsine), the carbyne substituent has an important influence on the product: only in the case of the methylcarbyne ligand is an additional bridging dimethylacetylene ligand formed, probably by coupling of two methylcarbyne ligands.

2 X(CO)$_4$W ≡ CMe + H$_2$C(AsPh$_2$)$_2$ →

[(CO)$_2$XW][(CO)$_3$W](μ-X)(μ-MeC ≡ CMe)(μ-Ph$_2$AsCH$_2$AsPh$_2$) + 3 CO

2 X(CO)$_4$W ≡ CPh + H$_2$C(AsPh$_2$)$_2$ →

[(CO)$_2$PhC ≡ W]$_2$(μ-X)$_2$(μ-Ph$_2$AsCH$_2$AsPh$_2$) + 4 CO

X = Cl, Br

On heating *trans*-tetracarbonyl(diethylaminocarbyne)(phenyltelluro)chromium in solution, a dimerisation with concomitant elimination of carbon monoxide takes place to give bis-(μ-phenyltelluro)-bis[tricarbonyl(diethylaminocarbyne)chromium] $(\mu\text{-TePh})_2[(CO)_3Cr \equiv CNEt_2]_2$ (62).

In a similar way, the intermediate mononuclear "tetracarbonylcarbyne complex" of manganese arises from the reaction of bromopentacarbonylmanganese with Buchner's mercuriodiazoalkanes, bis[(alkoxycarbonyl)diazomethyl]mercury $Hg(CN_2CO_2R)_2$. Herrmann has observed that the mononuclear intermediate undergoes a [2+2] cyclization to give a binuclear complex with two bridging carbyne ligands. In addition, this intermediate also yields a mononuclear carbyne/carbene coupling product $(CO)_4MnC(CO_2R) = C(HgBr)-C(OR) = O$ (63, 64).

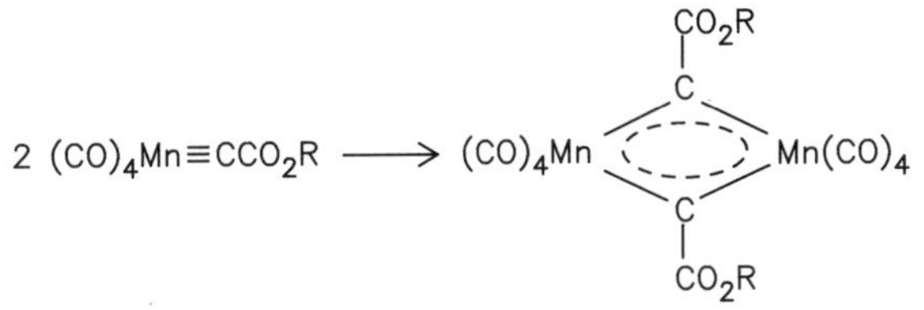

$$(CO)_4Mn \equiv CCO_2R + HgBr(CN_2CO_2R) \rightarrow$$

$$(CO)_4Mn[C(CO_2R) = C(HgBr)C(OR) = O] + N_2$$

R = Et, Bu^t

Roper has observed a similar [2+2] cyclization on the intermediate cationic osmium carbyne complex $[Cl(NO)(PPh_3)Os \equiv CCO_2Et]I$ (65).

In dicarbonyl(cyclopentadienyl)carbyne complexes of molybdenum and tungsten, the replacement of one carbonyl ligand occurs as a side reaction when these carbyne complexes are treated with trimethylphosphine (30, 66, 67).

$$Cp(CO)_2M \equiv CTol + PMe_3 \rightarrow Cp(CO)(PMe_3)M \equiv CTol + CO$$

M = Mo, W

2.4. Exchange of the Counter Ion

In cationic carbyne complexes, counter ion exchange is possible. Thus, tetrachloroborate salts can be converted with antimony pentachloride into the more stable $SbCl_6^-$ salts, and with silver tetrafluoroborate into the tetrafluoroborate salts (68).

$$[(CO)_5Cr \equiv CNR_2][BCl_4] + SbCl_5 \rightarrow [(CO)_5Cr \equiv CNR_2][SbCl_6] + BCl_3$$

This exchange reaction is often used to improve the thermal stability of cationic carbyne complexes and can also be applied to other cationic carbyne compounds.

3 Modification of the Carbyne Side Chain

Pentacarbonyl(diethylaminocarbyne)chromium tetrafluoroborate reacts with tetrabutylammonium azide to yield pentacarbonyl(diethylaminonitrile)-chromium instead of the expected pentacarbonyl(diethylamino(azido)carbene)chromium (69). It has been proposed that the first reaction step is the nucleophilic attack of azide at the carbyne carbon atom; this is followed by subsequent elimination of dinitrogen to give an intermediate diethylamino(nitrene)carbene compound, which could rearrange to the final reaction product.

$$[(CO)_5Cr \equiv CNEt_2]^+ + N_3^- \rightarrow \text{``}(CO)_5Cr = C(N_3)(NEt_2)\text{''} \rightarrow$$

$$\text{``}(CO)_5Cr = C(N)NEt_2\text{''} \rightarrow (CO)_5CrNCNEt_2$$

As in the case of the precursor carbene complexes, *trans*-halo(tetracarbonyl)-phenylethynylcarbynetungsten adds dimethylamine at $-25\,°C$ to the carbon-carbon triple bond to yield *trans*-halo(tetracarbonyl)(2-phenyl-2-dimethyl-aminoethylene)carbyne complexes (70).

$$X(CO)_4W \equiv CC \equiv CPh + HNMe_2 \rightarrow X(CO)_4W \equiv CC(H) = (Ph)NMe_2$$

X = Cl, Br, I

The stepwise incorporation of a carbyne ligand into vinylcarbene and vinylketene ligands has been reported for a tungsten carbyne complex with a special ligand combination (71). Thus, *trans*bromo(dicarbonyl)(phenylcarbyne)bis(4-picoline) tungsten reacts with allyl bromide with CC bond formation at the carbyne carbon to yield a vinylcarbene ligand.

$$Br(CO)_2(pic)_2W \equiv CPh + CH_2CHCH_2Br \rightarrow$$

$$Br_2(CO)_2(pic)W = C(Ph)CH = CHMe + \ldots$$

Successive addition of two equivalents of sodium diethyldithiocarbamate to the vinylcarbene complex gives an unstable intermediate dicarbonyl complex $(CO)_2(S_2CNEt_2)_2WC(Ph)CHCHMe$. Subsequent coupling of the vinylcarbene ligand with a carbonyl ligand results in the final formation of a vinylketene complex $CO(S_2CNEt_2)_2W(O = C = C(Ph)CHCHMe)$.

Upon addition of sodium hydride to a suspension of chlorobis(triisopropylphosphine)alkylcarbyneiridium tetrafluoroborate in equilibrium with chlorobis(triisopropylphosphine)alkenyliden(hydrido)iridium tetrafluoroborate, smooth evolution of hydrogen is observed and chlorobis(triisopropylphos-phine)alkenylideniridium is formed (72). (See also Chapter 8).

$$[Cl(P(Pr^i)_3)_2Ir \equiv CCH_2R][BF_4] + NaH \rightarrow$$

$$Cl(P(Pr^i)_3)_2Ir{=}C{=}CHR + H_2 + Na[BF_4]$$

R = H, Me, Ph

A further vinylidene complex is formed on the reaction of the acidic cationic carbyne complex [*mer*-(dppe)(CO)$_3$W $\equiv$ CCHMePh][BF$_4$] with 1,8-bis(dime-thylamino)naphthalene or alumina. Deprotonation of the cation yields the vin-ylidene moiety (47).

$$[mer\text{-}(dppe)(CO)_3W \equiv CCHMePh]^+ \rightarrow$$

$$mer\text{-}(dppe)(CO)_3W{=}C{=}CMePh + H^+$$

An unusual vinylidene complex results from hydride addition to the aromatic ring of the tolylcarbyne ligand in Roper's cationic carbyne complex [CO(PPh$_3$)$_2$L'Os $\equiv$ CTol]ClO$_4$ (L'=CO, CNR) (21).

$$[CO(PPh_3)_2L'Os{\equiv}CTol]^+ + H^- \longrightarrow CO(PPh_3)_2L'Os{=}C{=}C\underset{HC=CH}{\overset{HC=CH}{\diagdown}}CHMe$$

A very impressive modification of the carbyne side chain is represented by the nucleophilic displacement of the halide in HTPB(CO)$_2$M $\equiv$ CCl (M=Mo, W; HTPB=[HB(N$_2$C$_3$Me$_2$H)$_3$]$^-$, hydrotris(3,5-dimethylpyrazolyl)borate). This carbyne complex reacts readily with organothiolate and organoselenolate anions under phase-transfer conditions to yield the organothiocarbyne- and organoselenocarbyne complexes (73).

$$HTPB(CO)_2Mo \equiv CCl + Nu^- \rightarrow HTPB(CO)_2Mo \equiv CNu + Cl^-$$

Nu = SMe, SPh, SC$_6$H$_4$NO$_2$-**p**, SePh

A similar exchange of the chlorosubstituent by a tolylanion is postulated in the reaction of carbonyl(dichloro)(dichlorocarbene)bis(triphenylphosphine) osmium with tolyl lithium. The intermediate chlorocarbyne complex is suppos-ed to react *via* nucleophilic displacement of the halide (74).

$$Cl(CO)(PPh_3)_2Os \equiv CCl + LiTol \rightarrow Cl(CO)(PPh_3)_2Os \equiv CTol + LiCl$$

The chlorocarbyne complex $HTPB(CO)_2Mo \equiv CCl$ reacts with lithium sulfide, lithium selenide or sodium telluride in tetrahydrofuran or methanol to give anionic chalcogenocarbonyl complexes (73).

$$HTPB(CO)_2Mo \equiv CCl + Nu^{2-} \rightarrow [HTPB(CO)_2MoCNu]^- + Cl^-$$

Nu = S, Se, Te

The corresponding hydrotris(1-pyrazolyl)boratethiocarbynetungsten complexes can be synthesized by two different routes, but with lower overall yields, starting from thiocarbonyl compounds (75).

This method can also be applied in the nucleophilic displacement of the thiomethyl substituent in $HTPB(CO)_2W \equiv CSMe$ by trialkylphosphines to give cationic trimethylphosphinocarbyne complexes of tungsten (76). The SMe-group proved to be a very good leaving group in these displacement reactions.

$$HTPB(CO)_2W \equiv CSMe + PR_3 \rightarrow [HTPB(CO)_2W \equiv CPR_3]^+ + SMe^-$$

The reaction of the trimethylsilyl-substituted carbyne complexes of molybdenum with a solution of sodium fluoride in $MeCN/H_2O$ leads to desilylation and formation of the methyl- and benzyl-substituted carbyne complexes in high yields (77).

$$Ind[P(OMe)_3]_2Mo \equiv CCHRSiMe_3 \rightarrow Ind[P(OMe)_3]_2Mo \equiv CCH_2R + \ldots$$

R = H, Ph

The desilylation reaction can be explained in terms of nucleophilic attack by the fluoride anion at silicon with displacement of the anion $[Ind-[P(OMe)_3]_2Mo \equiv CCHR]^-$, which is then protonated at the β-carbon atom forming the corresponding carbyne complex.

4 Reactions with Nucleophiles

Experimental results, supported by molecular orbital calculations, may be considered as proof of the remarkable susceptibility of the carbyne carbon in transition metal carbyne complexes to nucleophilic attack.

4.1 Reactions of Carbyne Complexes with Phosphines

Strong basic neutral nucleophiles such as tertiary phosphines, especially trimethylphosphine, add under mild conditions to the carbyne carbon in transition metal carbyne complexes of chromium, molybdenum and tungsten. This reaction is closely related to the addition of phosphines to the carbene carbon in the precursor pentacarbonyl(aryl- or alkylalkoxy)carbene complexes and leads to metal-substituted phosphorus ylides or phosphinocarbene complexes (40, 43, 78–80),

$$X(CO)_4Cr \equiv CR + PMe_3 \rightarrow X(CO)_4Cr = C(R)PMe_3$$

M = Cr; R = Me, CH_2Ph, Ph, Tol, Mes, $SiPh_3$,
X = Cl, Br, I

In the case of molybdenum and tungsten as transition metals, the additional replacement of one or two carbonyl ligands takes place (40, 78, 79).

$$X(CO)_4M \equiv CR + 2 PMe_3 \longrightarrow X(CO)_3(PMe_3)M = C(R)PMe_3 + CO$$

$$\xrightarrow{PMe_3} X(CO)_2(PMe_3)_2W = C(R)PMe_3 + CO$$

M = Mo; R = Ph; X = Cl
M = W; R = Ph, $SiPh_3$; X = Cl, Br

This phosphine addition to the carbyne carbon has been extended, without further CO-substitution, to related cationic transition metal carbyne complexes of chromium, manganese and rhenium (81, 82).

$$[Ar(CO)_2M \equiv CPh][BCl_4] + PMe_3 \rightarrow [Ar(CO)_2M = C(Ph)PMe_3][BCl_4]$$

M = Cr; Ar = C_6H_6, 1,4-$Me_2C_6H_4$, 1,3,5-$Me_3C_6H_3$
M = Mn, Re; Ar = Cp

Closely related to these ylide reactions is the addition of pyridine to the carbyne carbon in dicarbonyl(cyclopentadienyl)phenylcarbynemanganese-tetrafluoroborate (83).

$$[Cp(CO)_2Mn \equiv CPh][BCl_4] + py \rightarrow [Cp(CO)_2Mn = C(Ph)py][BCl_4]$$

An X-ray crystallographic analysis indicated a pronounced Re-$C_{carbene}$ double bond (197(1) pm) and a $C_{carbene}$-P single bond (179(1) pm) for the rhenium phosphinocarbene complex $[Cp(CO)_2Re = C(Ph)PMe_3][BCl_4]$, supporting the formulation of the bond situation as a carbene rather than a ylide (84).

With isonitrile the manganese or rhenium complexes react to give first a comparable green cationic carbene complex (85, 86).

$$[Cp(CO)_2M \equiv CR][BCl_4] + CNR' \rightarrow [Cp(CO)_2M = C(R)CNR'][BCl_4]$$

M = Mn, Re; R = Me, Ph; R' = Me, c-C_6H_{11}, Bu^t

On thermolysis of the complex (R = Bu^t) the butyl group is split off to yield the neutral dicarbonyl(cyclopentadienyl)cyano(phenyl)carbenemanganese.

$$[Cp(CO)_2Mn = C(Ph)CNBu^t][BCl_4] \rightarrow Cp(CO)_2Mn = C(Ph)CN + \ldots$$

An interesting formation of a bis(phosphonium)carbene derivative takes place when $[HTPB(CO)_2W \equiv CPMe_3][PF_6]$ (HTPB = $[HB(N_2C_3Me_2H)_3]^-$, hydrotris(3,5-dimethylpyrazolyl)borate) reacts with trimethylphosphine (76).

$$[HTPB(CO)_2W \equiv CPMe_3][PF_6] + PMe_3 \rightarrow [HTPB(CO)_2W = C(PMe_3)_2][PF_6]$$

With an excess of trimethylphosphine, the cationic dicarbonylcyclopentadienylcarbyne complexes of manganese and rhenium react, at dry ice temperature, *via* intermediate metal-substituted ylides to yield thermolabile semi-ylide complexes (87, 88).

$$[Cp(CO)_2M \equiv CR][BF_4] + 2 PMe_3 \rightarrow [Cp(CO)_2M-C(R)(PMe_3)_2][BF_4]$$

M = Mn, Re; R = Ph, Tol, $SiPh_3$.

At room temperature, in the presence of trimethylphosphine, the manganese complex undergoes subsequent heterolytic cleavage of the metal carbon single bond to afford semi-ylide salts,

$$[Cp(CO)_2Mn\text{-}C(R)(PMe_3)_2][BF_4] + PMe_3 \rightarrow$$

$$[R\text{-}C(PMe_3)_2][BF_4] + Cp(CO)_2MnPMe_3$$

whereas, the rhenium semi-ylide compound loses one trimethylphosphine substituent to give the metal-substituted ylide.

$$[Cp(CO)_2Re\text{-}C(R)(PMe_3)_2][BF_4] \rightarrow$$

$$[Cp(CO)_2Re=C(R)(PMe_3)][BF_4] + PMe_3$$

In the case of *trans*-chloro(tetracarbonyl)phenylcarbynechromium the semi-ylide salt is formed in one step.

$$Cl(CO)_4Cr \equiv CPh + 2\ PMe_3 \rightarrow [Ph\text{-}C(PMe_3)_2]Cl + \ldots$$

The cationic pentacarbonyl(diethylaminocarbyne)chromium tetrafluoroborate displays a similar reaction behavior. In the presence of an excess of trimethylphosphine, the corresponding diethylamino-substituted semi-ylide salt is formed, at –40 °C, in very high yields (89).

$$[(CO)_5Cr \equiv CNEt_2][BF_4] + 3\ PMe_3 \rightarrow$$

$$[Et_2N\text{-}C(PMe_3)_2][BF_4] + (CO)_5CrPMe_3$$

However, pentacarbonyl(dimethylaminocarbyne)chromium-tetrafluoroborate and triethylphosphine react with displacement of the *trans*-carbonyl ligand by PEt₃ (53).

$$[(CO)_5Cr \equiv CNMe_2][BF_4] + PEt_3 \rightarrow [(CO)_4(PEt_3)Cr \equiv CNMe_2][BF_4] + CO$$

In a similar way, binuclear carbyne complexes like pentacarbonylrheniotetracarbonyl(arylcarbyne)tungsten add trimethylphosphine under mild conditions to form an ylide-bridged product (90, 91).

$$(CO)_5Re\text{-}(CO)_4M\equiv CPh + PMe_3 \longrightarrow (CO)_4Re\overset{Ph-C-PMe_3}{\underset{CO}{<\!\!>}}M(CO)_4$$

M = Cr, W; R = Ph, Tol

Depending on the nature of the aryl group, an additional displacement of one carbonyl ligand at the tungsten or the rhenium center is observed (91). This addition-rearrangement reaction demonstrates that, under special conditions, a bridging double-coordination of an ylide to transition metals is preferred over the terminal coordination.

4.2 Reactions of Cationic Carbyne Complexes with Nucleophiles

The pronounced electrophilic character of the carbyne carbon in the cationic carbyne complexes makes possible an important synthetic route to neutral carbene complexes. Compounds become available which are inaccessible *via* the usual synthetic approach.

Thus, substituted benzene(dicarbonyl)phenylcarbynechromium tetrafluoroborate and ammonia or dimethylamine form stable aminocarbene complexes (92).

$$[Ar(CO)_2Cr \equiv CPh][BCl_4] + HNR'_2 \rightarrow Ar(CO)_2Cr = C(Ph)NR'_2 + \ldots$$

$Ar = C_6H_6, MeC_6H_5, 1,4-Me_2C_6H_4, 1,3,5-Me_3C_6H_3; R' = H, Me.$

This reaction turned out to be quite general and could easily be applied to other cationic carbyne complexes of manganese, rhenium, chromium, molybdenum and tungsten.

In a similar way the $[BF_4]^-$, $[BCl_4]^-$ and $[SbCl_6]^-$ salts of dicarbonyl(cyclopentadienyl)phenylcarbyne manganese or rhenium add nucleophiles such as thiocyanate (93), cyanide (93), cyanate (94), fluoride (95, 96), chloride (96), bromide (96), iodide (96), alcoholates or phenolates (97), thio-, seleno- and tellurophenolates (98), and tetracarbonylcobaltate (99).

$$[Ar(CO)_2M \equiv CR]^+ + Nu^- \rightarrow Ar(CO)_2M = C(R)Nu$$

$M = Mn, Re; R = Ph, C_6H_4CF_3, C_5H_4FeC_5H_5;$
$Ar = Cp, MeCp; Nu = F, Cl, Br, I, CN, OCN, SCN, OR,$
$SPh, SePh, TePh, Co(CO)_4$

Treatment of the cationic carbyne complex $[MeCp(CO)_2Mn \equiv CPh][BF_4]$ with $LiC(=S)NMe_2$ yields the compound $MeCp(CO)_2Mn(S=C(Ph)SCH_2C(=S)NMe_2)$. This reaction not only involves the coupling of two thioformaldehyde units, the formation of a thiocarbene ligand by addition to the carbyne carbon atom, and a sulfur insertion into this metal-carbene bond, but also a very rare reductive deamination (100).

However, when the acylmetallate $[Cp(CO)_2Mn\text{-}C(=O)Ph]^-$ is used as a nucleophile, the following carbene anhydride complex is formed (101).

$$[Cp(CO)_2Mn \equiv CPh]^+ + [Cp(CO)_2Mn\text{-}C(=O)Ph]^- \rightarrow$$
$$[Cp(CO)_2Mn = C(Ph)]_2O$$

The reaction of methyllithium and lithium cyclopentadienyl with the cationic carbyne complexes of manganese yielded the first dimethylcarbene complex (102, 103) and a diorganylcarbene complex (94).

$$[Cp(CO)_2Mn \equiv CMe][BCl_4] + LiMe \rightarrow Cp(CO)_2Mn = CMe_2 + \ldots$$

$$[MeCp(CO)_2Mn \equiv CPh][BCl_4] + LiCp \rightarrow$$
$$MeCp(CO)_2Mn = C(Ph)C_5H_5 + \ldots$$

When $[Cp(CO)_2Re \equiv CSiPh_3][BCl_4]$ reacts with ethanol or dimethylamine two different addition products can be isolated. A similar reaction is reported for $[(CO)_5Re\text{-}(CO)_3Re \equiv CSiPh_3][BCl_4]$ (104, 105).

$$[Cp(CO)_2Re \equiv CSiPh_3][BCl_4] + HX \rightarrow$$
$$Cp(CQ)_2Re = C(X)SiPh_3 + Cp(CO)_2Re = C(X)H + \ldots$$

HX = EtOH, Me$_2$NH

Secondary carbene complexes are formed from the reaction of the rhenium complex with diethylaluminum hydride (106).

$$[Cp(CO)_2Re \equiv CPh][BCl_4] + Et_2AlH \rightarrow Cp(CO)_2Re = C(H)Ph + \ldots$$

The homologous tungsten complex $[HTPB(CO)_2W \equiv CPMe_3][PF_6]$ (76) (HTPB = $[HB(N_2C_3Me_2H)_3]^-$, hydrotris(3,5-dimethylpyrazolyl)borate) reacts with methyllithium or $K[HB(OPr^i)_3]$ in a similar way to give a secondary or a methyl(trimethylphosphino)carbene complex.

$$[HTPB(CO)_2W \equiv CPMe_3][PF_6] + MR \rightarrow HTPB(CO)_2W = C(PMe_3)R + \ldots$$

R = H, Me; M = Li, KB(OPr^i)$_3$

According to MO calculations the LUMO, in cationic carbyne complexes of the type $[(CO)_5Cr \equiv CNR_2]X$ is mainly localized on the carbyne carbon atom

(107). Consequently, these cations $[(CO)_5M \equiv CNR_2]X$ (M = Cr (53, 68, 108, 109), Mo (55), W (56, 58)) can be regarded as excellent precursors for the synthesis of numerous carbene complexes which are usually inaccessible *via* conventional direct routes. They react, even at low temperatures, with a huge variety of nucleophiles (F$^-$ (108, 109), Cl$^-$ (56, 110), Br$^-$ (111), I$^-$ (111), CN$^-$ (53, 112), NCO$^-$ (111), NCS$^-$ (111), NCSe$^-$ (112), SPh$^-$ (113), SeR$^-$ (57, 114), TePh$^-$ (62), AsPh$_2^-$ (115), SnPh$_3^-$ (116), PbPh$_3^-$ (117), $[(CO)_5MC(O)R]^-$ (118) (M = Cr, W; R = Me, Tol) and NPh$_2^-$ (119)) to form the corresponding carbene complexes.

$$[(CO)_5M \equiv CNR_2]^+ + Nu^- \rightarrow (CO)_5M = C(Nu)NR_2$$

M = Cr, Mo, W; Nu = F, Cl, Br, I, CN, NCO, NCS,
NCSe, SPh, SeR, TePh, AsPh$_2$, SnPh$_3$, PbPh$_3$,
$[(CO)_5MC(O)R]$, NPh$_2$

If these cationic carbyne complexes $[(CO)_5M \equiv CNR_2][BX_4]$ are allowed to react with strongly reducing nucleophiles such as R$_2$E$^-$ (E = P, As, Sb, Bi; R = Me, Et, Ph), completely different types of products are formed. Instead of the expected addition of the nucleophile to the carbyne carbon, a reductive C_aC_a-coupling of two carbyne fragments "$(CO)_5MCNEt_2$" takes place to give μ-bis-(aminocarbene) complexes (119, 120).

$$2 \ [(CO)_5M{\equiv}CNEt_2][BF_4] + 2 \ M'ER_2 \longrightarrow$$

$$(CO)_5M{=}C \overset{\textstyle NEt_2}{\underset{\textstyle C=M(CO)_5}{\diagdown}} \quad + \ R_2E{-}ER_2$$
$$\underset{Et_2N}{}$$

M = Cr, W; M' = Na, K;
E = P, As, Sb, Bi; R = Me, Et, Ph

With very strongly reducing nucleophiles, this competitive reductive dimerisation dominates over the addition of the nucleophile to the carbyne carbon.

5 Transfer of the Carbyne Ligand

In a few cases, the transfer of the carbyne ligand from one metal to another was achieved. Certain *trans*-halotetracarbonyl(organylcarbyne)complexes of chromium, molybdenum and tungsten, or dicarbonyl(cyclopentadienyl)phenyl-carbynemanganese tetrachloroborate react with octacarbonyldicobalt under very mild conditions to give π-alkyne-bis(tricarbonylcobalt) (121) and/or the well known μ³-alkylidyne(enneacarbonyl)tricobalt complexes (121, 122). Dimerisation of the carbyne ligands to alkynes and the subsequent formation of alkynehexacarbonyldicobalt complexes to form the μ³-alkylidyne-(enneacarbonyl)tricobalt complexes was ruled out.

$$X(CO)_4M \equiv CR + Co_2(CO)_8 \rightarrow$$

$$(CO)_9Co_3(\mu^3\text{-}CR) + (CO)_6Co_2(\pi\text{-}RC \equiv CR) + \ldots$$

$$[Cp(CO)_2Mn \equiv CPh][BCl_4] + Co_2(CO)_8 \rightarrow (CO)_9Co_3(\mu^3\text{-}CPh) + \ldots$$

M = Cr, Mo, W; X = Cl, Br; R = Me, Ph, Tol.

Under the same conditions, the reaction of bis(π-cyclopentadienyl)nickel with *trans*-bromotetracarbonyl(phenylcarbyne)chromium yields π-diphenyl-ethyne-bis(π-cyclopentadienylnickel). In addition, traces of μ³-benzylidyne-tris(π-cyclopentadienylnickel) were identified by means of mass spectrometry (121).

$$Br(CO)_4Cr \equiv CPh + NiCp_2 \rightarrow$$

$$(CpNi)_2(\pi\text{-}PhC \equiv CPh) + (CpNi)_3(\mu^3\text{-}CPh) + \ldots$$

At elevated temperatures, *trans*-bromo(tetracarbonyl)(organylcarbyne)-complexes of chromium react to yield alkynes (123).

$$Br(CO)_4Cr \equiv CR \rightarrow RC \equiv CR + \ldots$$

R = Me, Ph

Phenyltolylacetylene, and both symmetrical alkynes, tolane and bistolyl-acetylene, are formed on warming a mixture of *trans*-bromo(tetracarbonyl)phenylcarbyne chromium and *trans*-bromo(tetracarbonyl)tolylcarbyne chromium (123).

$$Br(CO)_4Cr \equiv CPh + Br(CO)_4Cr \equiv CTol \rightarrow$$

$$PhC \equiv CTol + PhC \equiv CPh + TolC \equiv CTol + \ldots$$

The mass spectra of the *trans*-halo(tetracarbonyl)organylcarbyne complexes of chromium display signals due to the disubstituted alkynes arising from the thermal decomposition of the carbyne complex and dimerisation of the carbyne ligands. Differences in the thermal decomposition rate of the neutral molecules suggest the following stability sequence: alkylcarbynechromium < arylcarbynechromium << organylcarbynetungsten (124).

trans-Halo(tetracarbonyl)carbyne complexes react with anions such as $C_6H_5^-$, OH^-, OR^- to yield carboxylic acids, carboxylic esters and ketones (125). The attack of the nucleophile at the positively charged carbonyl carbon atom is probably followed by the transfer of the resulting "acylate group" onto the carbyne carbon atom. Finally, protonation by acid initiates the reductive elimination of the entire organic ligand.

$$Br(CO)_4M \equiv CR + R'^- \rightarrow R\text{-}CH_2\text{-}CO\text{-}R' + \ldots$$

R = Ph, Tol; R' = OH, OEt, Ph

(*S*)-Phenyl-2-phenyl-2-(phenylthio)thioacetate is formed on reaction of *trans*-bromo(tetracarbonyl)-phenylcarbynechromium with lithium thiophenolate and subsequent protonation with hydrochloric acid (126).

$$Br(CO)_4Cr \equiv CPh + 2\ LiSPh \rightarrow Ph\text{-}CH(SPh)\text{-}COSPh + \ldots$$

This result can be explained in terms of an attack by the thiophenolate ions on both the carbonyl and the carbyne carbon atoms, and differs from the above described reaction with RO^-. The analogous reaction with selenophenolates does not lead to the corresponding ester, but to (diphenyldiselenide)pentacarbonylchromium and di(μ-phenylseleno)tetracarbonylchromium (126, 127).

trans-Bromotetracarbonylcarbyne complexes of tungsten induce alkynes and cycloalkenes to polymerize. The alkynes include examples that are unsubstituted, monosubstituted, and disubstituted as well as functionalized alkynes in which the functional groups (nitrile, ester, halogen) are not attached to the triple bond (128).

$$R'C \equiv CH \quad \xrightarrow{\quad Br(CO)_4W \equiv CR \quad} \quad polymers$$

R = Me, Ph

On the other hand, *trans*-bromo(tetracarbonyl)methylcarbynechromium or -tungsten combine stoichiometrically with a variety of alkynes to give specific phenols what seems to be a general process (129).

6 Base Induced Carbon-Carbon Coupling Reaction – Ketene Reaction

Dicarbonyl(η^5-cyclopentadienyl)carbyne complexes of molybdenum and tungsten react with strong bases such as phosphines, methylisonitrile or carbon monoxide *via* an intramolecular carbon-carbon coupling to yield η^1-and η^2-ketenyl complexes (130-138) and trimethylphosphine-substituted carbyne complexes.

$$Cp(CO)_2M \equiv CR + PR_3' \longrightarrow Cp(CO)(PR_3')M \overset{\displaystyle \overset{CO}{\diagup \diagdown}}{=\!\!=} CR$$

$$Cp(CO)_2M \equiv CR + 2\,PR_3' \longrightarrow Cp(CO)(PR_3')_2M - C \overset{\displaystyle C = O}{\underset{R}{\diagdown}}$$

M = Mo, W; R = Me, C_3H_5, C_5H_7-(1), Ph, Tol, Anis, Fc, Mes, $SiPh_3$; PR_3' = PMe_3, PEt_3, PPr_3, PBu_3, PPh_3, $P(OMe)_3$, $PMe_2(CH_2Ph)$, PMe_2Ph, $PMePh_2$

Similarly, Angelici's thiocarbyne complexes $[HB(pz)_3](CO)_2W \equiv CSR$, where $HB(pz)_3^-$ is the hydrotris(1-pyrazolylborato) ligand, react with phosphines to give the air-stable η^2-ketenyl complexes $[HB(pz)_3](CO)(PEt_3)$-$\overline{W\text{-}C(SR)}\overset{|}{C}O$ (139).

$$[HB(pz)_3](CO)_2W \equiv CSR + PR_3' \longrightarrow$$

$$[HB(pz)_3](CO)(PR'_3)W \overset{\displaystyle \overset{CO}{\diagup \diagdown}}{=\!\!=} C - SR$$

R = Me, DNP (= 2,4-dinitrophenyl);
PR_3' = PEt_3, PPh_3, $PMePh_2$

On reacting dicarbonyl(cyclopentadienyl)tolylcarbynetungsten with chelating phosphines such as $Me_2P(CH_2)_nPMe_2$ (n = 1, 2) or $(PF_2)_2NMe$ η^1-ketenyl compounds are formed (140).

$$Cp(CO)_2W\equiv CTol + Me_2P(CH_2)_nPMe_2 \longrightarrow$$

$$Cp(CO)(Me_2P(CH_2)_nPMe_2)W-C\underset{Tol}{\overset{C=O}{\diagdown}}$$

$$Cp(CO)_2W\equiv CTol + 2\ (PF_2)_2NMe \longrightarrow$$

$$Cp[(PF_2)_2NMe]_2W-C\underset{Tol}{\overset{C=O}{\diagdown}} + CO$$

Addition of one equivalent of sodium diethyldithiocarbamate to a solution of the cationic carbyne complex, $[(dppe)(CO)_2W\equiv CCH_2Ph][BF_4]$, yields first the neutral carbyne complex $(S_2CNEt_2)(dppe)(CO)_2W\equiv CCH_2Ph$. Subsequent chelation of the dithiocarbamate ligand promotes coupling of the carbyne ligand and carbon monoxide to form a η^2-ketenyl complex (47).

$$[(dppe)(CO)_2W\equiv CCH_2Ph]^+ + S_2CNMe_2^- \longrightarrow$$

$$S_2CNMe_2(dppe)(CO)_2W\equiv CCH_2Ph$$

$$S_2CNMe_2(dppe)(CO)_2W\equiv CCH_2Ph \longrightarrow$$

$$S_2CNMe_2(dppe)(CO)W\overset{CO}{\triangle}C-CH_2Ph$$

Anionic ketenyl complexes are formed from the reaction of neutral dicarbonyl(cyclopentadienyl)carbyne complexes of molybdenum and tungsten with tetraalkylammonium cyanide (141, 142)

$$Cp(CO)_2M\equiv CR + [NR_4']CN \longrightarrow$$

$$[NR_4'][Cp(CO)(CN)M\overset{CO}{\triangle}CR]$$

M = Mo, W; R = Me, Tol; R' = Et, Bu^n

or by treatment of $Br(CO)_2LL'W\equiv CR$ complexes with potassium cyanide (143, 144).

$$Br(CO)_2LL'W\equiv CR + 2\ KCN \longrightarrow$$

$$K[LL'(CO)(CN)_2W\overset{\overset{\displaystyle CO}{\diagup\diagdown}}{=\!\!=}CR] + KBr$$

LL' = phen, bipy; R = Me, Ph

Finally, Mayr's bis(pyridine)-substituted tungsten carbyne complexes react in tetrahydrofuran with two equivalents of pyrrole-2-carboxaldehyde methylimine (= HNN) in the presence of solid potassium hydroxide to give anionic tungsten ketenyl complexes which can be isolated as their tetraethylammonium salts (145). The analogous reaction of the phenylcarbyne complex $Cl(CO)_2(py)_2$-$W\equiv CPh$ with sodium diethyldithiocarbamate results in the formation of a further anionic ketenyl complex (146).

$$Cl(CO)_2(py)_2W\equiv CR + 2\ LL' + [NEt_4]^+ \longrightarrow$$

$$[NEt_4][CO(LL')_2W\overset{\overset{\displaystyle CO}{\diagup\diagdown}}{=\!\!=}CR] + ...$$

R = Me, Ph; LL' = NN, S_2CNEt_2

Only a few examples of reactions of carbon monoxide with carbene and carbyne complexes, such as Herrmann's high pressure carbonylation of diphenylcarbene(dicarbonyl)manganese (147), have so far been reported in the literature. If, however, carbonyl(cyclopentadienyl)tolylcarbyne(trimethylphosphine) complexes of molybdenum and tungsten (67) are allowed to react with carbon monoxide, carbonylation at the carbyne carbon takes place readily at normal pressure and –30 °C (148).

$$Cp(CO)(PMe_3)M\equiv CTol + 2\ CO \longrightarrow Cp(CO)_2(PMe_3)M-C\overset{\displaystyle C\diagup\!\!\diagup^{O}}{\diagdown_{Tol}}$$

M = Mo, W

A further CC coupling reaction has been reported as the result of UV irradiation of a solution of dicarbonyl(cyclopentadienyl)tolylcarbynetungsten. In the presence of carbon monoxide, carbonylation of the carbyne ligand and formation of a binuclear complex $(Cp(CO)_2W)_2(\mu\text{-}C(Tol)C(O)CTol)$ is induced (149). The bridging ligand of this compound consists of two carbyne units linked by carbon monoxide.

$$2\ Cp(CO)_2W\equiv CTol\ +\ CO\ \longrightarrow\ Cp(CO)_2W\underset{}{\overset{\displaystyle Tol-C\diagup\overset{CO}{\diagdown}C-Tol}{\diagup\diagdown}}W(CO)_2Cp$$

Upon repeating this irradiation in the presence of triphenylphosphine, formation of the earlier described ketenyl complexes $Cp(CO)(PPh_3)W(\eta^2\text{-}C(Tol)CO)$ and $Cp(CO)_2(PPh_3)W\text{-}C(Tol)CO$ is observed.

The cyclopentadienyl-substituted carbyne complexes of tungsten react with methylisonitrile to yield η^1-ketenyl complexes (135, 150).

A further interesting carbonyl carbyne coupling reaction is observed in the photochemical reaction between *trans*-chlorotetracarbonyl(tolylcarbyne)tungsten and acetylacetone which yields the complex-stabilized hydroxy(tolyl)acetylene ligand (151).

$$Cl(CO)_4W\equiv CTol\ +\ Hacac\ \rightarrow$$

$$\textit{trans-}Cl(CO)_2(acac)W(HO\text{-}C\equiv CTol)\ +\ CO$$

Treatment of bromo(tricarbonyl)(triphenylphosphine)phenylcarbynetungsten with lithium methyl and oxalyl bromide leads to an intermediate tungsten carbyne carbene complex which, upon dissociation of the bromooxalate, generates a new electron deficient methylcarbyne ligand. Coupling of the resulting two carbyne ligands occurs soon after, or possibly even simultaneously with, the dissociation of bromooxalate from the carbene ligand to form a methylphenylacetylene complex (151).

$$Br(CO)_3(PPh_3)W\equiv CPh\ +\ LiMe\ \longrightarrow$$

$$Li[Br(CO)_2(PPh_3)W(\equiv CPh)(-COMe)]$$

$$\Big\downarrow\ +\ C_2O_2Br_2$$

$$Br(CO)_2(PPh_3)W(\equiv CPh)(=C(Me)OC_2O_2Br$$

$$\Big\downarrow\ +\ PPh_3$$

$$Br_2(CO)(PPh_3)_2W(MeC\equiv CPh)$$

Binuclear ketenyl complexes with a metal-metal bond and a bridging ketene unit are formed by reacting mononuclear cationic carbyne complexes with carbonylmetallates. The transfer of a carbonyl ligand to the carbyne carbon gives rise to the formation of the new μ-C(R')=C=O moiety (99, 153–155).

$$[Ar(CO)_2M\equiv CR]^+ + [M'(CO)_5]^- \longrightarrow \quad Ar(CO)M\underset{CO}{\overset{R-C=C=O}{\diamondsuit}}M'(CO)_4$$

$$\longrightarrow$$

M = Mn, Re; M' = Mn, Re; Ar = Cp, MeCp;
R = Ph, Tol, Fc

$$[(CO)_5W\equiv CNEt_2]^+ + [Mn(CO)_5]^- \longrightarrow (CO)_4W\underset{CO}{\overset{Et_2N-C=C=O}{\diamondsuit}}Mn(CO)_4$$

These bridging ketenyl moieties can be described by a two-electron three-center system, as in Casey's cationic iron complexes (156).

$$Cp(CO)Fe\underset{CO}{\overset{\overset{\oplus}{C}-H}{\diamondsuit}}Fe(CO)Cp + CO \longrightarrow Cp(CO)Fe\underset{CO}{\overset{H-C=C=O}{\diamondsuit}}Fe(CO)Cp \quad \oplus$$

7 Electrophilic Addition of the Metal Carbon Triple Bond

Fenske's theoretical study of the cationic cyclopentadienyl-substituted phenylcarbyne complexes of manganese, $[Cp(CO)_2Mn \equiv CPh][BF_4]$, indicates that there is a build up of charge on the carbyne carbon atom (157). Thus, it is proposed that the formation of carbene complexes by nucleophilic attack on carbyne complexes is frontier orbital controlled. For the isoelectronic cyclopentadienylbis(trimethylphosphite)pentylcarbynemolybdenum, $Cp[P(OMe)_3]_2Mo \equiv CCH_2Bu^t$, a similar charge separation was found by extended Hückel M. O. calculations with a net charge on molybdenum and on the carbyne carbon atom of $+0.93$ and -0.34 e respectively (158). This result should be valid, in general, for other dicarbonyl(cyclopentadienyl)carbyne complexes of molybdenum and tungsten.

7.1 Protonation of Carbyne Complexes

The successful protonation of neutral carbyne complexes could be either frontier orbital or charge controlled, resulting, in the former case, in an electrophilic attack on the metal and in the latter, in a reaction at the carbyne carbon atom. Protonation of the molybdenum carbyne complex $Cp[P(OMe)_3]_2Mo \equiv CCH_2 Bu^t$ with $H[BF_4]$ leads to a hydrido-carbyne complex (158, 159), while a protolytic cleavage of the carbyne fragment is observed with trifluoroacetic acid to yield $Cp[P(OMe)_3]_2Mo(OOCCF_3)(O=C(OH)CF_3)$ (158).

$$Cp[P(OMe)_3]_2Mo\equiv CCH_2Bu^t + H[BF_4] \longrightarrow$$

$$[Cp[P(OMe)_3]_2(H)Mo\equiv CCH_2Bu^t][BF_4]$$

$$Cp[P(OMe)_3]_2Mo\equiv CCH_2Bu^t + n\ CF_3COOH \longrightarrow$$

$$Cp[P(OMe)_3]_2Mo \underset{O=C(OH)CF_3}{\overset{O-CO-CF_3}{\big<}}$$

In contrast, $ClL_4W \equiv CR$ (R = H, Bu^t) reacts with CF_3SO_3H to give hydrocarbene complexes, when L is trimethylphosphine. However, a hydridocarbyne complex is observed in the case of $Cl(dmpe)_2W \equiv CH$ (dmpe$= Me_2PCH_2CH_2PMe_2$) (160, 161).

$$Cl(PMe_3)_4W \equiv CR + CF_3SO_3H \rightarrow [Cl(PMe_3)_4W = CHR][CF_3SO_3]$$

R = H, Bu^t

$$Cl(dmpe)_2W \equiv CH + CF_3SO_3H \rightarrow [Cl(dmpe)_2(H)W \equiv CH][CF_3SO_3]$$

Schrock's tungsten carbyne complexes $(Bu^tO)_3W \equiv CR$ (R= Bu^t, Me, Et) show a distinctive reaction behavior upon protonation (162). Thus, $(Bu^tO)_3W \equiv CBu^t$ reacts with various acids to give carbene complexes.

$$(Bu^tO)_3W \equiv CBu^t + 2\ HX \rightarrow (Bu^tO)_2X_2W = C(H)Bu^t + Bu^tOH$$

X = Cl, Br, MeCO$_2$, PhCO$_2$, PhO, OC$_6$F$_5$, O-**p**-C$_6$H$_4$Cl

The corresponding methyl and ethylcarbyne complexes yield, with one equivalent of acetic acid, the carbene complexes. However, addition of two equivalents of RCOOH to $(Bu^tO)_3W \equiv CEt$ yields a propene complex (162). It has been proposed that the double protonation is followed by cleavage of a proton from the β-carbon of the intermediate propyl ligand.

$$(Bu^tO)_3W \equiv CEt + 2\ RCOOH \rightarrow (Bu^tO)_2(RCOO)_2W(H_2C = CHMe)$$

R = Me, Ph

The distinct protonation of the carbyne carbon is reported by Roper for an osmium carbyne complex, $CO(PPh_3)_2ClOs \equiv CTol$ (74, 163).

$$CO(PPh_3)_2(Cl)Os \equiv CTol + HX \rightarrow CO(PPh_3)_2(Cl)XOs = C(H)Tol$$

X = Cl, ClO$_4$

Further electrophilic attacks by protons have been described for the tungsten carbyne complexes $[HB(pz)_3](CO)_2W \equiv CSMe$ (139, 164) and $Cp(CO)_2W \equiv CNEt_2$ (165, 166). The first reacts to give the corresponding dihapto C- and S-coordinated mercaptocarbene complexes.

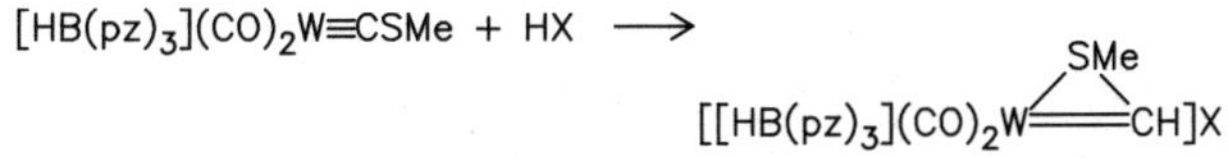

$$[HB(pz)_3](CO)_2W \equiv CSMe + HX \longrightarrow$$

$$[[HB(pz)_3](CO)_2W \overset{SMe}{=\!=\!=} CH]X$$

HX = CF$_3$COOH, H[BF$_4$] · Et$_2$O, CF$_3$COOH

Addition of the bases, sodium hydride or triethylamine, to the mercapto-carbene complex results in the regeneration of Angelici's carbyne complex. In the case of $Cp(CO)_2W \equiv CNEt_2$ diethylaminocarbene complexes are formed.

$$Cp(CO)_2W \equiv CNEt_2 + HX \rightarrow Cp(CO)_2XW = C(H)NEt_2$$

HX = HCl, CF$_3$COOH

In contrast, the dicationic methylaminocarbyne tungsten complex *trans*-$[(dppe)_2W(\equiv CNHMe)_2][BF_4]_2$ is protonated at the nitrogen (167).

$$trans\text{-}[(dppe)_2W(\equiv CNHMe)_2][BF_4]_2 + H[BF_4] \rightarrow$$
$$[(dppe)_2W(\equiv CNHMe)(\equiv CNH_2Me)][BF_4]_3$$

Deprotonation of the amido ligand and concomitant protonation of the carbyne carbon in $Cl_2(PEt_3)_2(NHPh)W \equiv CBu^t$ occurs during thermolysis (168).

$$Cl_2(PEt_3)_2(NHPh)W \equiv CBu^t \rightarrow Cl_2(PEt_3)_2W(= C(H)Bu^t)(= NPh)$$

The reactions of the dicarbonyl(cyclopentadienyl)organylcarbyne complexes of molybdenum and tungsten with protic nucleophiles lead to different products. Protonation of dicarbonyl(cyclopentadienyl)organylcarbyne tungsten with $H[BF_4] \cdot Et_2O$ affords the ditungsten compounds $[Cp_2(CO)_4W_2(\mu\text{-H})(\mu\text{-}RC \equiv CR)][BF_4]$, formed *via* coupling of the carbyne ligands in the precursors (169, 170).

$$2\ Cp(CO)_2W \equiv CR + H[BF_4] \cdot Et_2O \rightarrow$$
$$[Cp(CO)_2W(\mu\text{-H})(\mu\text{-}RC \equiv CR)W(CO)_2Cp][BF_4] + \ldots$$

R = Me, Tol

Treatment of these carbyne complexes with aqueous hydrogen iodide affords the corresponding iodo carbene complexes (170).

$$Cp(CO)_2W \equiv CR + HI \rightarrow Cp(CO)_2IW = C(H)R$$

On the contrary, protonation of the dicarbonyl(cyclopentadienyl)organylcarbyne complexes of molybdenum and tungsten with the protic nucleophiles, hydrogen chloride, trifluoroacetic acid or trichloroacetic acid, yields dihapto acyl complexes in high yields (171, 172).

$$Cp(CO)LM\equiv CR + 2\ HX \longrightarrow Cp(L)X_2M\underset{O}{\overset{}{\diagup}}C-CH_2R$$

M = Mo, W; R = Me, Ph, Tol, C_3H_5; L = CO, PMe_3;
X = Cl, CF_3COO, CCl_3COO

It has been proposed that these carbyne-acyl rearrangements start with the protonation of the carbyne carbon atom to give a carbene complex. For the next step, two different pathways have been discussed: (i). An intramolecular carbene carbonyl coupling to yield a π-ketene complex which undergoes a further protonation to give the final product. (ii). Subsequent protonation of the intermediate carbene carbon atom to form an alkyl compound, which undergoes a pseudocarbonyl insertion reaction yielding the η^2-acyl compound (173).

$$Cp(CO)_2X_2M-CH_2R$$

$$Cp(CO)_2XM=C(H)R \qquad\qquad Cp(CO)X_2M(\eta^2\text{-}O=C-CH_2R)$$

$$Cp(CO)XM(\pi\text{-}O=C=C(H)R)$$

7.2 Addition of Electrophiles to the Metal Carbon Triple Bond

In a novel reaction sequence, characterized by a twofold electrophilic attack on the carbon atom of a metal carbon multiple bond, certain dicarbonyl(cyclopentadienyl)organylcarbyne complexes of molybdenum and tungsten react in a stepwise, easily observed manner with dimethyl(methylthio)sulfonium tetrafluoroborate to give first the cationic η^2-thiocarbene complex and then the dicationic dithiatungstabicyclo[1.1.0]butane complex (173, 174).

$$Cp(CO)_2M\equiv CR + [Me_2(MeS)S]^+ \longrightarrow [Cp(CO)_2M\overset{SMe}{\diagdown\!=\!\!=\!\diagup}CR)]^+ + \ldots$$

$$\downarrow + SMe^+$$

$$[Cp(CO)_2M\overset{SMe}{\underset{SMe}{\diagdown\!-\!\diagup}}CR)]^{2+}$$

M = Mo, W; R = Me, Ph, Tol, NEt_2

With halo phosphines or halo arsines, dicarbonyl(cyclopentadienyl)carbyne complexes of tungsten yield thermolabile η^3-ketene complexes (175).

$$Cp(CO)_2W\equiv CR + XER_2' \longrightarrow Cp(CO)(X)W\cdots\begin{array}{c} R \\ R_2E-C \\ \backslash\backslash \\ C \\ \backslash\backslash \\ O \end{array}$$

R = Me, Ph, Tol; R' = Me, Ph; X = Cl, I; E = P, As

Treatment of dicarbonyl(cyclopentadienyl)tolylcarbynetungsten with diphenylphosphine or lithium diphenylphosphide/ammonium bromide gives the cyclic compounds $Cp(CO)_2W$-C(H)(Tol)-PPh$_2$ and $Cp(CO)(PHPh_2)$-WC(H)(Tol)-CO-O-PPh$_2$ (176).

$$Cp(CO)_2W\equiv CTol + HPPh_2 \longrightarrow$$

$$Cp(CO)_2\overline{W-C(H)(Tol)-PPh_2} +$$

$$Cp(CO)(PHPh_2)\overline{WC(H)(Tol)-CO-O-PPh_2}$$

Reaction of certain cyclopentadienyl-substituted carbyne complexes, $CpL_2Mo\equiv CCH_2Bu^t$ (L = CO, P(OMe)$_3$) and $Cp(CO)_2W\equiv CTol$, with sulphur or selenium affords complexes with a metalla-(dithia or diselena)-cyclobutane moiety (177). It has been suggested that the initial step of these reactions is the nucleophilic attack of an S$_8$ or Se$_8$ chain at the carbyne carbon to give first an "S$_8$- or Se$_8$-carbene" ligand.

$$CpL_2M\equiv CR + X_8 \longrightarrow CpL_2M\begin{array}{c} X \\ \diagup\ \diagdown \\ \diagdown\ \diagup \\ X \end{array}C-R$$

M = Mo, W; L = CO, P(OMe)$_3$; X = S, Se

Treatment of dicarbonyl(cyclopentadienyl)methylcarbynetungsten with an excess of BH$_3 \cdot$ thf at room temperature results in the formation of an interesting dinuclear tungsten-boron complex (178). The ethyl group attached to boron must result from cleavage of the tungsten carbyne triple bond. All three hydrogens of the BH$_3$ group are incorporated into the final product, two have been transferred to the former carbyne carbon and the third forms a tungsten boron hydride bridge!

$$2 \; Cp(CO)_2W \equiv CMe + BH_3 \longrightarrow Cp(CO)_2W \overset{\displaystyle CMe}{\underset{\displaystyle Et-B-H}{\Big<\Big|\Big>}} W(CO)_2Cp$$

In contrast, dicarbonyl(cyclopentadienyl)tolylcarbynetungsten reacts readily with 9-borabicyclo[3.3.1]nonane (= 9-BBN) to form a mononuclear complex with an unusual η^3-ligand (178).

$$Cp(CO)_2W \equiv CTol + BBN \rightarrow Cp(CO)_2W[\eta^3\text{-Tol-C(H)BC}_8H_{14}]$$

A similar η^3-tolyl(phenyl)benzyl complex can be obtained upon treatment of dicarbonyl(cyclopentadienyl)tolylcarbynetungsten with organo chromium or organo zinc compounds (179).

$$Cp(CO)_2W \equiv CTol + CrPh_3(thf)_3 \rightarrow$$
$$Cp(CO)_2W[\eta^3\text{-Tol-C(H)Ph}] + \ldots$$

$$Cp(CO)_2W \equiv CTol + ZnR_2' \rightarrow Cp(CO)_2W[\eta^3\text{-Tol-C(H)R'}] + \ldots$$

R' = Pr^i, $CH_2C_6H_4Me$

However, diethylzinc affords a trinuclear metal complex $Zn[Cp(CO)_2W(\eta\text{-}MeCH = CHTol)]_2$, which, in the presence of traces of water, yields the η^3-tolyl-(phenyl)benzyl complex $Cp(CO)_2W[\eta^3\text{-Tol-C(H)Et}]$.

7.3 Reactions of the Carbyne Complexes with Metal Complexes

The osmium carbyne complex $Cl(CO)(PPh_3)_2Os \equiv CTol$ reacts with halides of the coinage metals to form mixed dimetallic η^2-carbene or bridging carbyne complexes (74, 180). A single X-ray structure determination for the AgCl-adduct confirms the dimetallacyclopropene moiety.

$$Cl(CO)(PPh_3)_2Os \equiv CTol + MX \longrightarrow Cl(CO)(PPh_3)_2Os \overset{\displaystyle \overset{X}{\underset{\displaystyle M}{|}}}{\diagdown}\!\!=\!\!= CTol$$

MX = CuI, AgCl, AgClO$_4$, AuCl(PPh$_3$)

The synthetic potential of carbyne complexes for the preparation of di-, tri- and polynuclear transition metal complexes has been impressively demonstrated by F. G. A. Stone (181–183). The Stone reaction connects mainly dicarbonyl(cyclopentadienyl)carbyne complexes of tungsten with other metal complexes to synthesize a wide variety of new interesting complexes. It utilizes Hoffmann's (184) isolobal relationship between the carbon-carbon triple bond and the metal-carbon triple bond. Thus, bisphosphine(ethene)platinum reacts with dicarbonyl-substituted tungsten carbyne complexes *via* addition at the tungsten carbon triple bond to yield bridging carbyne complexes (36, 37, 185, 186).

$$Ar(CO)_2W{\equiv}CTol + L_2Pt(C_2H_4) \longrightarrow Ar(CO)_2W{=\!=}CTol \overset{\displaystyle PtL_2}{\diagup\diagdown}$$

Ar = Cp, HBpz$_3$; L = PMe$_2$Ph

There are many other complexes with transition metals in addition to these platinum compounds. Titanium (187, 188), zirconium (187), vanadium (189), chromium (190), molybdenum (28, 191–194), tungsten (192–196), manganese (197), rhenium (197, 198), iron (28, 34, 199–204), ruthenium (205–209), osmium (205), cobalt (28, 34, 197, 210, 211), rhodium (36, 197, 202, 211–214), iridium (197), nickel (34, 212, 215–218), palladium (215, 217), platinum (216–219), copper (220), silver (193) and gold (193, 214, 221) react readily with cyclopentadienyl-, pentamethylcyclopentadienyl- or carbaborane-substituted carbyne complexes to give interesting di-, tri- and polynuclear metal complexes with bridging carbene or carbyne ligands. Similar mixed-metal complexes are formed upon treatment of the binuclear carbyne complexes $(CO)_nM'$-$(CO)_4M{\equiv}CR$ with different transition metal complexes (222–226).

8 Oxidation and Reduction Reactions

Oxidation and reduction reactions can be carried out on both the metal and the metal-carbyne bond (8, 227). The Fischer-type carbyne complex *trans*-bromo(tetracarbonyl)phenylcarbynechromium reacts with chlorine and bromine as well as with cerium(IV) and manganese(III) compounds with complete degradation of the complex. Benzotrihalides or benzoic acid derivatives and, in some cases, also dibenzyl or tolane are obtained (228).

$$Br(CO)_4Cr \equiv CPh + n\ X_2 \rightarrow PhCX_3 + \ldots$$

X = Cl, Br

$$Br(CO)_4Cr \equiv CPh + Ce(SO_4)_2 \cdot 4\ H_2O \rightarrow Ph\text{-}C \equiv C\text{-}Ph + \ldots$$

$$Br(CO)_4Cr \equiv CPh + [NH_4]_2[Ce(NO_3)_6] \rightarrow Ph\text{-}COOMe + \ldots$$

$$Br(CO)_4Cr \equiv CPh + [Mn_3OAc_6]Ac \rightarrow Ph\text{-}COOMe + Ph\text{-}CH_2\text{-}CH_2\text{-}Ph + \ldots$$

The first step in these oxidation reactions is proposed to be the formation of a radical cation *via* oxidation of the central metal (229).

$$Br(CO)_4\ Cr \equiv CPh \rightarrow [(Br(CO)_4Cr \equiv CPh)]^+ + e^-$$

Addition of one equivalent of chlorine to chloro(carbonyl)bis(triphenylphosphine)tolylcarbyneosmium results in the formation of a chlorocarbene complex (74, 163).

$$Cl(CO)(PPh_3)_2Os \equiv CTol + Cl_2 \rightarrow Cl_2(CO)(PPh_3)_2Os = C(Cl)Tol$$

Reaction of this carbyne complex with elemental sulfur, selenium or tellurium yields dihapto(chalcogeno)acyl complexes (74, 163).

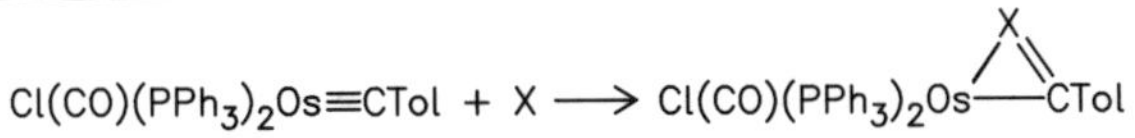

$$Cl(CO)(PPh_3)_2Os \equiv CTol + X \longrightarrow Cl(CO)(PPh_3)_2Os \overset{\displaystyle X}{\underset{}{\diagup\!\!\diagdown}} CTol$$

X = S, Se, Te

When chloro(carbonyl)bis(triphenylphosphine)phenylcarbyneosmium is exposed to oxygen at 0 °C, the only isolable product, in high yields, is chloro-(dicarbonyl)(phenyl)bis(triphenylphosphine)osmium (230).

$$Cl(CO)(PPh_3)_2Os \equiv CPh + O_2 \rightarrow Cl(CO)_2(PPh_3)_2(Ph)Os + \ldots$$

In contrast, the zero-valent p-dimethylaminophenylcarbyne complex of osmium, $Cl(CO)(PPh_3)_2Os \equiv CC_6H_4NMe_2$, reacts with molecular oxygen to yield a yellow, crystalline 1:1 adduct. This has been formulated as a divalent, octahedral complex retaining the unchanged carbyne ligand, and with a dihapto-peroxocarbonyl ligand (230).

$$Cl(CO)(PPh_3)_2Os \equiv CC_6H_4NMe_2 + O_2 \longrightarrow Cl(PPh_3)_2Os \equiv CC_6H_4NMe_2$$

The reaction of one equivalent of [PPN][NO$_2$] (PPN = (Ph$_3$P)$_2$N) with cyclopentadienyl(dicarbonyl)tolylcarbynerhenium tetrachloroborate results in the formal addition of O$^-$ to the carbyne ligand forming an acyl ligand, followed by substitution of NO for carbon monoxide (231).

$$[Cp(CO)_2Re \equiv CTol][BCl_4] + NO_2^- \rightarrow Cp(CO)(NO)Re\text{-}C(O)Tol + CO + \ldots$$

However, upon repeating this reaction with the homologous manganese compound, [MeCp(CO)$_2$Mn $\equiv$ CR][BCl$_4$], a novel arylglyoxyl complex was formed (231), involving the formal coupling of an acyl and a carbonyl ligand.

$$[MeCp(CO)_2Mn \equiv CR][BCl_4] + NO_2^- \rightarrow MeCp(CO)(NO)Mn\text{-}\underset{\displaystyle \overset{R\text{-}C=O}{\diagdown}}{C}=O + \ldots$$

R = Ph, Tol

In both reactions, the initial step is supposed to be the nucleophilic attack of the anion NO$_2^-$ on the carbyne carbon.

A typical example of the reduction of a carbyne complex can be seen in the reaction of an anionic rhenium carbyne complex with diethylaluminum hydride (106) (eg, "see Chapter 4.2"). Depending on the reaction conditions, the secondary carbene complex, Cp(CO)$_2$Re = CHPh, or the σ-benzylhydrido complex, Cp(CO)$_2$(H)Re-CH$_2$Ph, is formed.

$$[Cp(CO)_2Re \equiv CPh]^+ + Et_2AlH \nearrow \begin{array}{l} Cp(CO)_2Re=CHPh + \ldots \\ \\ Cp(CO)_2(H)Re\text{-}CH_2Ph + \ldots \end{array}$$

9 References

(1) E. O. Fischer, G. Kreis, C. G. Kreiter, J. Müller, G. Huttner, H. Lorenz, *Angew. Chem. Int. Ed. Engl.* (1973) **12** 564–565.

(2) E. O. Fischer, U. Schubert, H. Fischer, *Pure & Appl. Chem.* (1978) **50** 857–870.

(3) U. Schubert, in: *The Chemistry of Metal-Carbon Bond:* F. R. Hertley, S. Patai (eds.): John Wiley & Sons Ltd (1982); pp. 233–243.

(4) E. O. Fischer, U. Schubert, *J. Organomet. Chem.* (1975) **100** 59–81.

(5) R. R. Schrock, *J. Am. Chem. Soc.* (1974) **96** 6796–6797.

(6) D. N. Clark, R. R. Schrock, *J. Am. Chem. Soc.* (1978) **100** 6774–6776.

(7) E. O. Fischer, T. L. Lindner, F. R. Kreißl, P. Braunstein, *Chem. Ber.* (1977) **110** 3139–3148.

(8) E. O. Fischer, M. Schluge, J. O. Besenhard, P. Friedrich, G. Huttner, F. R. Kreißl, *Chem. Ber.* (1978) **111** 3530–3541.

(9) E. O. Fischer, W. R. Wagner, F. R. Kreißl, D. Neugebauer, *Chem. Ber.* (1979) **112** 1320–1328.

(10) E. O. Fischer, H. Hollfelder, F. R. Kreißl, *Chem. Ber.* (1979) **112** 2177–2189.

(11) H. Fischer, F. Seitz, *J. Organomet. Chem.* (1984) **268** 247–258.

(12) H. Fischer, F. Seitz, *J. Organomet. Chem.* (1984) **275** 83–91.

(13) E. O. Fischer, A. C. Filippou, H. G. Alt, *J. Organomet. Chem.* (1985) **296** 69–82.

(14) A. C. Filippou, E. O. Fischer, K. Öfele, H. G. Alt, *J. Organomet. Chem.* (1986) **308** 11–17.

(15) A. C. Filippou, E. O. Fischer, H. G. Alt, U. Thewalt, *J. Organomet. Chem.* (1987) **326** 59–81.

(16) E. O. Fischer, K. Weiß, C. G. Kreiter, *Chem. Ber.* (1974) **107** 3554–3561.

(17) E. O. Fischer, K. Weiß, *Chem. Ber.* (1976) **109** 1128–1139.

(18) K. Richter, E. O. Fischer, C. G. Kreiter, *J. Organomet. Chem.* (1976) **122** 187–196.

(19) E. O. Fischer, S. Walz, A. Ruhs, F. R. Kreißl, *Chem. Ber.* (1978) **111** 2765–2773.

(20) E. O. Fischer, F. J. Gammel, *Z. Naturforsch.* (1979) **34B** 1183–1185.

(21) W. R. Roper, J. M. Waters, L. J. Wright, *J. Organomet. Chem.* (1980) **201** C27–C30.

(22) E. O. Fischer, N. H. Tran-Huy, D. Neugebauer, *J. Organomet. Chem.* (1982) **229** 169–177.

(23) E. O. Fischer, J. K. R. Wanner, *J. Organomet. Chem.* (1983) **252** 175–179.

(24) E. O. Fischer, G. Huttner, T. L. Lindner, A. Frank, F. R. Kreißl, *Angew. Chem. Int. Ed. Engl.* (1976) **15** 157.

(25) E. O. Fischer, T. L. Lindner, F. R. Kreißl, P. Braunstein, *Chem. Ber.* (1977) **110** 3139–3148.

(26) S. Fontana, O. Orama, E. O. Fischer, U. Schubert, F. R. Kreißl, *J. Organomet. Chem.* (1978) **149** C57–C62.

(27) E. O. Fischer, P. Friedrich, T. L. Lindner, D. Neugebauer, F. R. Kreißl, W. Uedelhoven, *J. Organomet. Chem.* (1983) **247** 239–246.

(28) M. D. Bermudez, E. Delgado, G. P. Elliott, N. H. Tran-Huy, F. Mayor-Real, F. G. A. Stone, M. J. Winter, *J. Chem. Soc. Dalton Trans.* (1987) 1235–1242.

(29) F. R. Kreißl, W. Uedelhoven, K. Eberl, unpublished results 1979.

(30) W. Uedelhoven, K. Eberl, F. R. Kreißl, *Chem. Ber.* (1979) **112** 3376–3389.

(31) W. J. Sieber, Dissertation Technische Universität München, 1984.

(32) E. O. Fischer, T. L. Lindner, F. R. Kreißl, *J. Organomet. Chem.* (1976) **112** C27–C30.

(33) E. O. Fischer, T. L. Lindner, G. Huttner, P. Friedrich, F. R. Kreißl, J. O. Besenhard, *Chem. Ber.* (1977) **110** 3397–3404.

(34) E. Delgado, J. Hein, J. C. Jeffery, A. L. Ratermann, F. G. A. Stone, L. J. Farrugia, *J. Chem. Soc. Dalton Trans.* (1987) 1191–1199.

(35) M. Green, J. A. K. Howard, A. P. James, A. N. de M. Jelfs, C. M. Nunn, F. G. A. Stone, *J. Chem. Soc. Chem. Commun.* (1984) 1623–1625.

(36) M. Green, J. A. K. Howard, A. P. James, C. M. Nunn, F. G. A. Stone, *J. Chem. Soc. Dalton Trans.* (1986) 187–197.

(37) M. Green, J. A. K. Howard, A. P. James, C. M. Nunn, F. G. A. Stone, *J. Chem. Soc. Chem. Commun.* (1984) 1113–1115.

(38) W. Kläui, H. Hamers, *J. Organomet. Chem.* (1988) **344** C27–C30.

(39) E. O. Fischer, A. C. Filippou, H. G. Alt, U. Thewalt, *Angew. Chem. Int. Ed. Engl.* (1985) **24** 203–205.

(40) E. O. Fischer, A. Ruhs, F. R. Kreißl, *Chem. Ber.* (1977) **110** 805–815.

(41) E. O. Fischer, K. Richter, *Chem. Ber.* (1976) **109** 2547–2557.

(42) A. Filippou, E. O. Fischer, *Z. Naturforsch.* (1983) **38B** 587–591.

(43) F. R. Kreißl, *J. Organomet. Chem.* (1975) **99** 305–308.

(44) A. Mayr, G. A. McDermott, A. M. Dorries, *Organometallics* (1985) **4** 608–610.

(45) G. A. McDermott, A. M. Dorries, A. Mayr, *Organometallics* (1987) **6** 925–931.

(46) A. Mayr, A. M. Dorries, G. A. McDermott, D. Van Engen, *Organometallics* (1986) **5** 1504–1506.

(47) K. R. Birdwhistell, T. L. Tonker, J. L. Templeton, *J. Am. Chem. Soc.* (1985) **107** 4474–4483.

(48) A. Mayr, A. M. Dorries, G. A. McDermott, *J. Am. Chem. Soc.* (1985) **107** 7775–7776.

(49) A. C. Filippou, E. O. Fischer, *J. Organomet. Chem.* (1986) **310** 357–366.

(50) A. C. Filippou, E. O. Fischer, *J. Organomet. Chem.* (1987) **330** C1–C4.

(51) E. O. Fischer, D. Wittmann, *J. Organomet. Chem.* (1985) **292** 245–246.

(52) H. Fischer, A. Motsch, U. Schubert, D. Neugebauer, *Angew. Chem. Int. Ed. Engl.* (1981) **20** 463–464.

(53) A. J. Hartshorn, M. F. Lappert, *J. Chem. Soc. Chem. Comm.* (1976) 761–762.

(54) H. Fischer, A. Motsch, R. Märkl, K. Ackermann, *Organometallics* (1985) **4** 726–735.

(55) E. O. Fischer, D. Wittmann, D. Himmelreich, R. Cai, K. Ackermann, D. Neugebauer, *Chem. Ber.* (1982) **115** 3152–3166.

(56) F. R. Kreißl, W. Uedelhoven, unpublished results.

(57) E. O. Fischer, D. Himmelreich, R. Cai, *Chem. Ber.* (1982) **115** 84–89.

(58) E. O. Fischer, D. Wittmann, D. Himmelreich, U. Schubert, K. Ackermann, *Chem. Ber.* (1982) **115** 3141–3151.

(59) E. O. Fischer, A. Ruhs, P. Friedrich, G. Huttner, *Angew. Chem. Int. Ed. Engl.* (1977) **16** 465–466.

(60) E. O. Fischer, A. Ruhs, *Chem. Ber.* (1978) **111** 2774–2778.

(61) E. O. Fischer, W. Kellerer, B. Zimmer-Gasser, U. Schubert, *J. Organomet. Chem.* (1980) **199** C24–C26.

(62) H. Fischer, E. O. Fischer, R. Cai, D. Himmelreich, *Chem. Ber.* (1983) **116** 1009–1016.

(63) W. A. Herrmann, *Angew. Chem. Int. Ed. Engl.* (1974) **13** 812.

(64) W. A. Herrmann, M. L. Ziegler, O. Serhadli, *Organometallics* (1983) **2** 958–962.

(65) M. A. Gallop, T. C. Jones, C. E. F. Rickard, W. R. Roper, *J. Chem. Soc. Chem. Comm.* (1984) 1002–1003.

(66) F. R. Kreißl, K. Eberl, W. Uedelhoven, *IXth International Conference on Organometallic Chemistry, Dijon, September 1979, Abstracts,* pp. B9.

(67) F. R. Kreißl, W. Uedelhoven, D. Neugebauer, *J. Organomet. Chem.* (1988) **344** C27–C30.

(68) R. Märkl, H. Fischer, *J. Organomet. Chem.* (1984) **267** 277–284.

(69) E. O. Fischer, W. Kleine, U. Schubert, D. Neugebauer, *J. Organomet. Chem.* 1978 **149** C40–C42.

(70) E. O. Fischer, H. J. Kalder, F. H. Köhler, *J. Organomet. Chem.* (1974) **81** C23–C27.

(71) A. Mayr, M. F. Asaro, T. J. Glines, *J. Am. Chem. Soc.* (1987) **109** 2215–2216.

(72) H. Werner, A. Höhn, *Angew. Chem. Int. Ed. Engl.* (1986) **25** 737–738.

(73) T. Desmond, F. J. Lalor, G. Ferguson, M. Parvez, *J. Chem. Soc. Chem. Commun.* (1984) 75–77.

(74) G. R. Clark, C. M. Cochrane, K. Marsden, W. R. Roper, L. J. Wright, *J. Organomet. Chem.* (1986) **315** 211–230.

(75) W.W. Greaves, R. J. Angelici, *Inorg. Chem.* (1981) **20** 2983–2988.

(76) A. E. Bruce, A. S. Gamble, T. L. Tonker, J. L. Templeton, *Organometallics* (1987) **6** 1350–1352.

(77) St. R. Allen, R. G. Beevor, M. Green, A. G. Orpen, K. E. Paddick, I. D. Williams, *J. Chem. Soc. Dalton Trans.* (1978) 591–604.

(78) F. R. Kreißl, W. Uedelhoven, A. Ruhs, *J. Organomet. Chem.* (176) **113** C55–C57.

(79) F. R. Kreißl, W. Uedelhoven, G. Kreis, *Chem. Ber.* (1978) **111** 3283–3293.

(80) H. Schmidbaur, *Angew. Chem. Int. Ed. Engl.* (1983) **22** 907–927.

(81) F. R. Kreißl, P. Stückler, *J. Organomet. Chem.* (1976) **110** C9–C11.

(82) F. R. Kreißl, P. Stückler, E.W. Meineke, *Chem. Ber.* (1977) **110** 3040–3045.

(83) E.W. Meineke, Dissertation Technische Universität München, 1975.

(84) F. R. Kreißl, P. Friedrich, *Angew. Chem. Int. Ed. Engl.* (1977) **16** 543–544.

(85) E. O. Fischer, W. Schambeck, F. R. Kreißl, *J. Organomet. Chem.* (1979) **169** C27–C30.

(86) E. O. Fischer, W. Schambeck, *J. Organomet. Chem.* (1980) **201** 311–318.

(87) F. R. Kreißl, K. Eberl, P. Stückler, *Angew. Chem. Int. Ed. Engl.* (1977) **16** 654.

(88) W. Uedelhoven, K. Eberl, W. Sieber, F. R. Kreißl, *J. Organomet. Chem.* (1982) **236** 301–307.

(89) F. R. Kreißl, K. Eberl, W. Kleine, *Chem. Ber.* (1978) **111** 2451–2452.

(90) F. R. Kreißl, P. Friedrich, T. L. Lindner, G. Huttner, *Angew. Chem. Int. Ed. Engl.* (1977) **16** 314.

(91) W. Uedelhoven, D. Neugebauer, F. R. Kreißl, *J. Organomet. Chem.* (1981) **217** 183–194.

(92) E. O. Fischer, P. Stückler, H.-J. Beck, F. R. Kreißl, *Chem. Ber.* (1976) **109** 3089–3098.

(93) E. O. Fischer, P. Stückler, F. R. Kreißl, *J. Organomet. Chem.* (1977) **129** 197–202.

(94) E. O. Fischer, G. Besl, *Z. Naturforsch.* (1979) **34B** 1186–1189.

(95) E. O. Fischer, W. Kleine, W. Schambeck, U. Schubert, *Z. Naturforsch.* (1981) **36B** 1575–1579.

(96) E. O. Fischer, J. Chen, K. Scherzer, *J. Organomet. Chem.* (1983) **253** 231–241.

(97) E. O. Fischer, E.W. Meineke, F. R. Kreißl, *Chem. Ber.* (1977) **110** 1140–1147.

(98) E. O. Fischer, J. K. R. Wanner, *Chem. Ber.* (1985) **118** 2489–2492.

(99) E. O. Fischer, J. K. R. Wanner, G. Müller, J. Riede, *Chem. Ber.* (1985) **118** 3311–3319.

(100) H. G. Raubenheimer, G. J. Kruger, A. van A. Lombard, L. Linford, J. C. Viljoen, *Organometallics* (1985) **4** 275–284.

(101) E. O. Fischer, J. Chen, U. Schubert, *Z. Naturforsch.* (1982) **37B** 1284–1288.

(102) E. O. Fischer, R. L. Clough, G. Besl, F. R. Kreißl, *Angew. Chem. Int. Ed. Engl.* (1976) **15** 543–544.

(103) E. O. Fischer, R. L. Clough, P. Stückler, *J. Organomet. Chem.* (1976) **120** C6–C8.

(104) E. O. Fischer, P. Rustemeyer, K. Ackermann, *Chem. Ber.* (1982) **115** 3851–3859.

(105) E. O. Fischer, P. Rustemeyer, *J. Organomet. Chem.* (1982) **225** 265–277.

(106) E. O. Fischer, A. Frank, *Chem. Ber.* (1978) **111** 3740–3744.

(107) U. Schubert, D. Neugebauer, P. Hofmann, B. E. R. Schilling, H. Fischer, A. Motsch, *Chem. Ber.* (1981) **114** 3349–3365.

(108) E. O. Fischer, W. Kleine, F. R. Kreißl, *Angew. Chem. Int. Ed. Engl.* (1976) **15** 616–617.

(109) E. O. Fischer, W. Kleine, G. Kreis, F. R. Kreißl, *Chem. Ber.* (1978) **111** 3542–3551.

(110) A. Motsch, Dissertation, Technische Universität München, 1980.

(111) E. O. Fischer, W. Kleine, F. R. Kreißl, H. Fischer, P. Friedrich, G. Huttner, *J. Organomet. Chem.* (1977) **128** C49–C53.

(112) W. Kleine, Dissertation, Technische Universität München, 1978.

(113) D. Wittmann, Dissertation, Technische Universität München, 1982.

(114) E. O. Fischer, D. Himmelreich, R. Cai, H. Fischer, U. Schubert, B. Zimmer-Gasser, *Chem. Ber.* (1981) **114** 3209–3219.

(115) U. Schubert, E. O. Fischer, D. Wittmann, *Angew. Chem. Int. Ed. Engl.* (1980) **19** 643–644.

(116) E. O. Fischer, R. B. A. Pardy, U. Schubert, *J. Organomet. Chem.* (1979) **181** 37–45.

(117) H. Fischer, E. O. Fischer, R. Cai, *Chem. Ber.* (1982) **115** 2707–2713.

(118) E. O. Fischer, W. Kleine, *J. Organomet. Chem.* (1981) **208** C27–C30.

(119) E. O. Fischer, R. Reitmeier, *Z. Naturforsch.* (1983) **B38** 582–586.

(120) E. O. Fischer, D. Wittmann, D. Himmelreich, D. Neugebauer, *Angew. Chem. Int. Ed. Engl.* (1982) **21** 444–445.

(121) E. O. Fischer, A. Däweritz, *Angew. Chem. Int. Ed. Engl.* (1975) **14** 346–347.

(122) E. O. Fischer, A. Däweritz, *Chem. Ber.* (1978) **111** 3525–3529.

(123) E. O. Fischer, A. Ruhs, D. Plabst, *Z. Naturforsch.* (1977) **B32** 802–804.

(124) W. Kalbfus, E. O. Fischer, J. W. Buchler, *J. Organomet. Chem.* (1977) **129** 79–90.

(125) E. O. Fischer, T. L. Lindner, *Z. Naturforsch.* (1977) **B32** 713–714.

(126) E. O. Fischer, W. Röll, *Angew. Chem. Int. Ed. Engl.* (1980) **19** 205–206.

(127) W. Röll, E. O. Fischer, D. Neugebauer, U. Schubert, *Z. Naturforsch.* (1982) **B37** 1274–1278.

(128) T. J. Katz, T. H. Ho, N.-Y. Shih, Y.-C. Ying, V. I.W. Stuart, *J. Am. Chem. Soc.* (1984) **106** 2659–2668.

(129) T. M. Sivavec, T. J. Katz, *Tetrahedron Letters* (1985) **26** 2159–2162.

(130) F. R. Kreißl, A. Frank, U. Schubert, T. L. Lindner, G. Huttner, *Angew. Chem. Int. Ed. Engl.* (1976) **15** 632–633.

(131) F. R. Kreißl, P. Friedrich, G. Huttner, *Angew. Chem. Int. Ed. Engl.* (1977) **16** 102–103.

(132) F. R. Kreißl, K. Eberl, W. Uedelhoven, *Chem. Ber.* (1977) **110** 3782–3791.

(133) F. R. Kreißl, in Organometallics in Organic Synthesis, edited by A. de Meijere and H. tom Dieck, Springer-Verlag, Berlin Heidelberg 1987, 105–119.

(134) W. Sieber, M. Wolfgruber, N. H. Tran-Huy, H. R. Schmidt, H. Heiß, P. Hofmann, F. R. Kreißl, *J. Organomet. Chem.* (1988) **340** 341–351.

(135) K. Eberl, Dissertation, Technische Universität München, 1979.

(136) F. R. Kreißl, in: *Organometallic Syntheses:* R. B. King, J. J. Eisch (eds.) Amsterdam: Elsevier Science Publishers (1986) Vol. 3, pp. 241–245.

(137) F. R. Kreißl, in: *Organometallic Syntheses:* R. B. King, J. J. Eisch (eds.) Amsterdam: Elsevier Science Publishers (1986) Vol. 3, pp. 246–249.

(138) J. C. Jeffery, C. Sambale, M. F. Schmidt, F. G. A. Stone, *Organometallics* (1982) **1** 1597–1604.

(139) H. P. Kim, S. Kim, R. A. Jacobson, R. J. Angelici, *Organometallics* (1986) **5** 2481–2488.

(140) K. Eberl, W. Uedelhoven, H. H. Karsch, F. R. Kreißl, *Chem. Ber.* (1980) **113** 3377–3380.

(141) W. J. Sieber, K. Eberl, M. Wolfgruber, F. R. Kreißl, *Z. Naturforsch.* (1983) **B38** 1159–1160.

(142) F. R. Kreißl, W. J. Sieber, H. G. Alt, *Chem. Ber.* (1984) **117** 2527–2530.

(143) E. O. Fischer, A. C. Filippou, H. G. Alt, K. Ackermann, *J. Organomet. Chem.* (1983) **254** C21–C23.

(144) E. O. Fischer, A. C. Filippou, H. G. Alt, *J. Organomet. Chem.* (1984) **276** 377–385.

(145) A. Mayr, G. A. McDermott, A. M. Dorries, D. van Engen, *Organometallics* (1987) **6** 1503–1508.

(146) A. Mayr, G. A. McDermott, A. M. Dorries, A. K. Holder, *J. Am. Chem. Soc.* (1986) **108** 310–311.

(147) W. A. Herrmann, J. Plank, *Angew. Chem. Int. Ed. Engl.* (1978) **17** 525–526.

(148) F. R. Kreißl, W. Uedelhoven, K. Eberl, *Angew. Chem. Int. Ed. Engl.* (1978) **17** 859.

(149) J. B. Sheridan, G. L. Geoffroy, A. L. Rheingold, *Organometallics* (1986) **5** 1514–1515.

(150) F. R. Kreißl, K. Eberl, W. Uedelhoven, *Xth International Conference on Organometallic Chemistry, Toronto, September 1981, Abstracts.*

(151) E. O. Fischer, P. Friedrich, *Angew. Chem. Int. Ed. Engl.* (1979) **18** 327–328.

(152) G. A. McDermott, A. Mayr, *J. Am. Chem. Soc.* (1987) **109** 580–582.

(153) O. Orama, U. Schubert, F. R. Kreißl, E. O. Fischer, *Z. Naturforsch.* (1980) **B35** 82–85.

(154) J. Martin-Gil, J. A. K. Howard, R. Navarro, F. G. A. Stone, *J. Chem. Soc. Chem. Comm.* (1979) 1168–1169.

(155) D. Himmelreich, Dissertation, Technische Universität München, 1982.

(156) C. P. Casey, P. J. Fagan, V. W. Day, *J. Am. Chem. Soc.* (1982) **104** 7360–7361.

(157) N. M. Kostic, R. F. Fenske, *J. Am. Chem. Soc.* (1981) **103** 4677–4685.

(158) M. Green, A. G. Orpen, I. D. Williams, *J. Chem. Soc. Chem. Commun.* (1982) 493–495.

(159) M. Bottrill, M. Green, *J. Am. Chem. Soc.* (1977) **99** 5795–5796.

(160) S. J. Holmes, R. R. Schrock, *J. Am. Chem. Soc.* (1981) **103** 4599–4600.

(161) S. J. Holmes, D. N. Clark, H. W. Turner, R. R. Schrock, *J. Am. Chem. Soc.* (1982) **104** 6322–6329.

(162) J. H. Freudenberger, R. R. Schrock, *Organometallics* (1985) **4** 1937–1944.

(163) G. R. Clark, K. Marsden, W. R. Roper, L. J. Wright, *J. Am. Chem. Soc.* (1980) **102** 6570–6571.

(164) H. P. Kim, S. Kim, R. A. Jacobson, R. J. Angelici, *Organometallics* (1984) **3** 1124–1126.

(165) F. R. Kreißl, W. J. Sieber, M. Wolfgruber, *J. Organomet. Chem.* (1984) **270** C45–C47.

(166) M. Wolfgruber, F. R. Kreißl, unpublished results.

(167) A. J. L. Pombeiro, R. L. Richards, *Transition Met. Chem.* (1980) **5** 55–59.

(168) S. M. Rocklage, R. R. Schrock, M. R. Churchill, H. J. Wasserman, *Organometallics* (1982) **1** 1332–1338.

(169) J. C. Jeffery, J. C. V. Laurie, I. Moore, F. G. A. Stone, *J. Organomet. Chem.* (1983) **258** C37–C40.

(170) J. A. K. Howard, J. C. Jeffery, J. C. V. Laurie, I. Moore, F. G. A. Stone, A. Stringer, *Inorganica Chimica Acta* (1985) **100** 23–32.

(171) F. R. Kreißl, W. J. Sieber, M. Wolfgruber, J. Riede, *Angew. Chem. Int. Ed. Engl.* (1984) **23** 640.

(172) F. R. Kreißl, W. J. Sieber, H. Keller, J. Riede, M. Wolfgruber, *J. Organomet. Chem.* (1987) **320** 83–90.

(173) F. R. Kreißl, H. Keller, N. Ullrich, unpublished results.

144   *Fritz R. Kreissl*

(174) F. R. Kreißl, H. Keller, *Angew. Chem. Int. Ed. Engl.* (1976) **25** 904.

(175) F. R. Kreißl, M. Wolfgruber, C. Stegmair, unpublished results.

(176) G. A. Carriedo, V. Riera, M. L. Rodriguez, J. C. Jeffery, *J. Organomet. Chem.* (1986) **314** 139–149.

(177) D. S. Gill, M. Green, K. Marsden, I. Moore, A. G. Orpen, F. G. A. Stone, I. D. Williams, P. Woodward, *J. Chem. Soc. Dalton. Trans.* (1984) 1343–1347.

(178) G. A. Carriedo, G. P. Elliott, J. A. K. Howard, D. B. Lewis, F. G. A. Stone, *J. Chem. Soc. Chem. Commun.* (1984) 1585–1586.

(179) J. C. Jeffery, A. L. Ratermann, F. G. A. Stone, *J. Organomet. Chem.* (1985) **289** 367–376.

(180) G. R. Clark, C. M. Cochrane, W. R. Roper, L. J. Wright, *J. Organomet. Chem.* (1980) **199** C35–C38.

(181) F. G. A. Stone, *Inorg. Chim. Acta,* (1981) **50** 33–42.

(182) F. G. A. Stone, *Angew. Chem. Int. Ed. Engl.* (1984) **23** 89–99.

(183) F. G. A. Stone, in: Inorganic Chemistry: towards the 21st Century: M. H. Chisholm (ed.). *Am. Chem. Soc., Symp. Ser.* (1983) **211** 383.

(184) R. Hoffmann, *Angew. Chem. Int. Ed. Engl.* (1982) **21** 711–800.

(185) T. V. Ashworth, J. A. K. Howard, F. G. A. Stone, *J. Chem. Soc. Chem. Commun.* (1979) 42–43.

(186) T. V. Ashworth, J. A. K. Howard, F. G. A. Stone, *J. Chem. Soc. Dalton. Trans.* (1980) 1609–1614.

(187) G. M. Dawkins, M. Green, K. A. Mead, J.-Y. Salaün, F. G. A. Stone, P. Woodward, *J. Chem. Soc. Dalton. Trans.* (1983) 527–530.

(188) M. R. Awang, R. D. Barr, M. Green, J. A. K. Howard, T. B. Marder, F. G. A. Stone, *J. Chem. Soc. Dalton. Trans.* (1985) 2009–2016.

(189) U. Behrens, F. G. A. Stone, *J. Chem. Soc. Dalton. Trans.* (1984) 1605–1607.

(190) J. C. Jeffery, J. C. V. Laurie, I. Moore, H. Razay, F. G. A. Stone, *J. Chem. Soc. Dalton. Trans.* (1984) 1563–1569.

(191) J. C. Jeffery, J. C. V. Laurie, F. G. A. Stone, *Polyhedron* (1985) **4** 1135–1139.

(192) M. Green, S. J. Porter, F. G. A. Stone, *J. Chem. Soc. Dalton. Trans.* (1983) 513–517.

(193) G. A. Carriedo, J. A. K. Howard, F. G. A. Stone, P. Woodward, *J. Chem. Soc. Dalton. Trans.* (1984) 1589–1595.

(194) M. Green, J. A. K. Howard, A. P. James, A. N. de M. Jelfs, C. M. Nunn, F. G. A. Stone, *J. Chem. Soc. Dalton. Trans.* (1987) 81–90.

(195) G. A. Carriedo, J. A. K. Howard, D. B. Lewis, G. E. Lewis, F. G. A. Stone, *J. Chem. Soc. Dalton. Trans.* (1985) 905–912.

(196) D. Hodgson, J. A. K. Howard, F. G. A. Stone, M. J. Went, *J. Chem. Soc. Dalton. Trans.* (1985) 1331–1337.

(197) J. A. Abad, L.W. Bateman, J. C. Jeffery, K. A. Mead, H. Razay, F. G. A. Stone, P. Woodward, *J. Chem. Soc. Dalton. Trans.* (1983) 2075–2081.

(198) G. A. Carriedo, J. C. Jeffery, F. G. A. Stone, *J. Chem. Soc. Dalton. Trans.* (1984) 1597–1603.

(199) L. Busetto, J. C. Jeffery, R. M. Mills, F. G. A. Stone, M. Went, P. Woodward, *J. Chem. Soc. Dalton. Trans.* (1983) 101–109.

(200) E. Delgado, A.T. Emo, J. C. Jeffery, N. D. Simmons, F. G. A. Stone, *J. Chem. Soc. Dalton. Trans.* (1985) 1323–1329.

(201) M. Green, J. A. K. Howard, A. P. James, A. N. de M. Jelfs, C. M. Nunn, F. G. A. Stone, *J. Chem. Soc. Dalton. Trans.* (1986) 1697–1707.

(202) S.V. Hoskins, A. P. James, J. C. Jeffery, F. G. A. Stone, *J. Chem. Soc. Dalton. Trans.* (1986) 1709–1716.

(203) E. Delgado, J. C. Jeffery, F. G. A. Stone, *J. Chem. Soc. Dalton. Trans.* (1986) 2105–2112.

(204) M. E. Garcia, J. C. Jeffery, P. Sherwood, F. G. A. Stone, *J. Chem. Soc. Dalton. Trans.* (1987) 1209–1214.

(205) L. Busetto, M. Green, B. Hessner, J. A. K. Howard, J. C. Jeffery, F. G. A. Stone, *J. Chem. Soc. Dalton. Trans.* (1983) 519–525.

(206) J. A. K. Howard, J. C.V. Laurie, O. Johnson, F. G. A. Stone, *J. Chem. Soc. Dalton. Trans.* (1985) 2017–2024.

(207) L. J. Farrugia, J. C. Jeffery, C. Marsden, P. Sherwood, F. G. A. Stone, *J. Chem. Soc. Dalton. Trans.* (1987) 51–59.

(208) M. Green, J. A. K. Howard, A. N. de M. Jelfs, O. Johnson, F. G. A. Stone, *J. Chem. Soc. Dalton. Trans.* (1987) 73–79.

(209) D. L. Davies, M. J. Parrott, P. Sherwood, F. G. A. Stone, *J. Chem. Soc. Dalton. Trans.* (1987) 1201–1208.

(210) M. J. Chetcuti, P. A. M. Chetcuti, J. C. Jeffery, R. M. Mills, P. Mitrprachachon, S. J. Pickering, F. G. A. Stone, P. Woodward, *J. Chem. Soc. Dalton. Trans.* (1982) 699–708.

(211) J. C. Jeffery, C. Marsden, F. G. A. Stone, *J. Chem. Soc. Dalton. Trans.* (1985) 1315–1321.

(212) M. Green, J. C. Jeffery, S. J. Porter, H. Razay, F. G. A. Stone, *J. Chem. Soc. Dalton. Trans.* (1982) 2475–2483.

(213) M. Green, J. A. K. Howard, S. J. Porter, F. G. A. Stone, D. C. Tyler, *J. Chem. Soc. Dalton. Trans.* (1984) 2553–2559.

(214) M. Green, J. A. K. Howard, A. P. James, C. M. Nunn, F. G. A. Stone, *J. Chem. Soc. Dalton. Trans.* (1987) 61–72.

(215) T.V. Ashworth, M. J. Chetcuti, J. A. K. Howard, F. G. A. Stone, S. J. Wisbey, P. Woodward, *J. Chem. Soc. Dalton. Trans.* (1981) 763–770.

(216) G. P. Elliott, J. A. K. Howard, T. Mise, I. Moore, C. M. Nunn, F. G. A. Stone, *J. Chem. Soc. Dalton. Trans.* (1986) 2091–2103.

(217) S. H. F. Becke, M. D. Bermudez, N. H. Tran-Huy, J. A. K. Howard, O. Johnson, F. G. A. Stone, *J. Chem. Soc. Dalton. Trans.* (1987) 1229–1234.

(218) S. J. Davies, G. P. Elliott, J. A. K. Howard, C. M. Nunn, F. G. A. Stone, *J. Chem. Soc. Dalton. Trans.* (1987) 2177–2187.

(219) M. R. Awang, J. C. Jeffery, F. G. A. Stone, *J. Chem. Soc. Dalton. Trans.* (1986) 165–172.

(220) G. A. Carriedo, J. A. K. Howard, F. G. A. Stone, *J. Chem. Soc. Dalton. Trans.* (1984) 1555–1561.

(221) G. A. Carriedo, J. A. K. Howard, F. G. A. Stone, M. J. Went, *J. Chem. Soc. Dalton. Trans.* (1984) 2545–2551.

(222) D. G. Evans, J. A. K. Howard, J. C. Jeffery, D. B. Lewis, G. E. Lewis, M. J. Grosse-Ophoff, M. J. Parrott, F. G. A. Stone, *J. Chem. Soc. Dalton. Trans.* (1986) 1723–1730.

(223) J. C. Jeffery, D. B. Lewis, G. E. Lewis, M. J. Parrott, F. G. A. Stone, *J. Chem. Soc. Dalton. Trans.* (1986) 1717–1722.

(224) J. A. Abad, E. Delgado, M. E. Garcia, M. J. Grosse-Ophoff, I. J. Hart, J. C. Jeffery, M. S. Simmons, F. G. A. Stone, *J. Chem. Soc. Dalton. Trans.* (1987) 41–50.

(225) J. C. Jeffery, A. G. Orpen, F. G. A. Stone, M. J. Went, *J. Chem. Soc. Dalton. Trans.* (1986) 173–186.

(226) J. C. Jeffery, D. B. Lewis, G. E. Lewis, F. G. A. Stone, *J. Chem. Soc. Dalton. Trans.* (1985) 2001–2007.

(227) E. O. Fischer, F. J. Gammel, J. O. Besenhard, A. Frank, D. Neugebauer, *J. Organomet. Chem.* (1980) **191** 261–282.

(228) E. O. Fischer, A. Ruhs, H.-J. Kalder, *Z. Naturforsch.* (1977) **B32** 473–475.

(229) H.-J. Kalder, Dissertation, Technische Universität München, 1976.

(230) G. R. Clark, N. R. Edmonds, R. A. Pauptit, W. R. Roper, J. M. Waters, A. H. Wright, *J. Organomet. Chem.* (1983) **244** C57–C60.

(231) J. B. Sheridan, G. L. Geoffroy, *J. Am. Chem. Soc.* (1987) **109** 1584–1586.

High Oxidation State
Alkylidyne Complexes

John S. Murdzek and Richard R. Schrock

1 Introduction

Organometallic compounds possessing a triple bond between the metal and an alkylidyne (CR) ligand comprise a large and varied class. Complexes in which the metal can be said to be in its highest possible oxidation state (d^0, counting the alkylidyne ligand as a trianion) can be separated conveniently on that basis from complexes in which the metal is not in its highest possible oxidation state (i.e., carbyne or Fischer-type complexes). Fischer-type complexes are reviewed elsewhere in this book. Our purpose in this chapter is to review d^0 alkylidyne complexes up to the end of 1987, including some data from our own program that have not been published elsewhere. A few selected lower oxidation state complexes are mentioned when they have been prepared from or yield d^0 species. A short review of molybdenum and tungsten d^0 alkylidyne complexes

Table 1. Structure Data for d^0 Alkylidyne Complexes.

Compound	$M{\equiv}C_\alpha$ (pm)	$M\text{-}C_\alpha\text{-}C_\beta$ (°)	Ref.
[Li(dmp)][Ta(CCMe$_3$)(CH$_2$CMe$_3$)$_3$]	176(2)	165(1)	(2)
Ta(CPh)(η^5-C$_5$Me$_5$)(PMe$_3$)$_2$Cl	184.9(8)	171.8(6)	(12)
Ta(CCMe$_3$)(H)(dmpe)$_2$(ClAlMe$_3$)	185.0(5)	178.7(4)	(18)
[Ta(μ-CCMe$_3$)Cl$_2$(dme)]$_2$[Zn(μ-Cl)$_2$]			
	186(2), 179(2)	156(1), 159(1)	(22)
[Nb(μ-CSiMe$_3$)(CH$_2$SiMe$_3$)$_2$]$_2$	199.5(9), 195.4(8)	119.8(6), 142.4(5)	(19)
Ta$_2$(μ-CSiMe$_3$)$_2$(CH$_2$SiMe$_3$)$_3$(O-2,6-C$_6$H$_3$Ph$_2$)			
	193(2), 196(1)	128.7(6), 134.5(6)	(21)
Ta$_2$(μ-CSiMe$_3$)$_2$(CH$_2$SiMe$_3$)$_2$(O-2,6-C$_6$H$_3$(CMe$_3$)$_2$)$_2$			
	199.7(6), 200.2(6)	122.1(3), 133.9(4)	(21)
W(CCMe$_3$)(PMe$_3$)$_3$Cl$_3$	179.3(6)	178.6(4)	(32)
W(CCMe$_3$)(CHCMe$_3$)(CH$_2$CMe$_3$)(dmpe)			
	178.5(8)	175.23(69)	(28)
[W(CMe)(OCMe$_3$)$_3$]$_2$	175.9(6)	179.8(6)	(42)
W(CPh)(OCMe$_3$)$_3$	175.8(5)	175.8(4)	(41)
[W(CNMe$_2$)(OCMe$_3$)$_3$]$_2$	177(2), 175(2)	179(2)	(54)
W[C-Ru(CO)$_2$(η^5-C$_5$H$_5$)](OCMe$_3$)$_3$	175(2)	177(2)	(51)
W(CCMe$_3$)(PHPh)Cl$_2$(PEt$_3$)$_2$	180.8(6)	174.0(4)	(82)
{W[η^5-C$_5$Me$_4$(CMe$_3$)](CCMe$_3$)I}$_2$(μ-N$_2$H$_2$)			
	176.9(8)	171.87(67)	(87)
Re(CCMe$_3$)(CHCMe$_3$)I$_2$(py)$_2$	174.2(9)	174.7(7)	(117)

has been published recently (1). The present review will cover Mo and W more extensively as well as Nb, Ta, and Re, the only other metals for which d^0 alkylidyne complexes are known. Structure data are given in Table 1. The known niobium and tantalum, molybdenum, tungsten, and rhenium complexes are listed in Tables 2-5, respectively.

Table 2. Niobium and Tantalum Alkylidyne Complexes.

Compound	C_α (ppm)	$^2J_{CP}$ (Hz)	Ref.
$[LiL_2][Ta(CCMe_3)(CH_2CMe_3)_3]^{(a)}$			(2)
$Ta(\eta^5\text{-}C_5H_5)(CCMe_3)(PMe_3)_2Cl$	348	16	(9)
$Ta(\eta^5\text{-}C_5H_5)(CPh)(PMe_3)_2Cl$	334	21	(9)
$Ta(\eta^5\text{-}C_5Me_5)(CCMe_3)(PMe_3)_2Cl$	354	19	(9)
$Ta(\eta^5\text{-}C_5Me_5)(CPh)(PMe_3)_2Cl$	345	20	(9)
$Ta(\eta^5\text{-}C_5Me_5)(CCMe_3)(H)(PMe_3)_2$			(11)
$Ta(\eta^5\text{-}C_5Me_5)(CCMe_3)(H)(dmpe)$	306.4	9	(11)
$Ta(CCMe_3)(H)(dmpe)_2(ClAlMe_3)$	272.3		(18)
$Ta(CCMe_3)(H)(dmpe)_2(I)$	248.3$^{(b)}$		(18)
$Ta(CCMe_3)(H)(dmpe)_2(O_3SCF_3)$	279.3		(18)
$[Ta(\mu\text{-}CCMe_3)Cl_2(dme)]_2[Zn(\mu\text{-}Cl)_2]$			(22)
$[Nb(\mu\text{-}CSiMe_3)(CH_2SiMe_3)_2]_2$	406		(20)
$[Ta(\mu\text{-}CSiMe_3)(CH_2SiMe_3)_2]_2$	406		(20)
$Ta_2(\mu\text{-}CSiMe_3)_2(CH_2SiMe_3)_3(O\text{-}2,6\text{-}C_6H_3Ph_2)$			
	399.3		(21)
$Ta_2(\mu\text{-}CSiMe_3)_2(CH_2SiMe_3)_2(O\text{-}2,6\text{-}C_6H_3Ph_2)_2$			(21)
$Ta_2(\mu\text{-}CSiMe_3)_2(CH_2SiMe_3)_2[O\text{-}2,6\text{-}C_6H_3(CMe_3)_2]_2$			
	386.3		(21)

(a) L_2 = N,N'-dimethylpiperazine, 2 THF, TMEDA, 2 p-dioxane, or 1,2-dimethoxyethane.

(b) This signal is an average of the C_α resonance for the neopentylidene and neopentylidyne complexes.

2 Niobium and Tantalum Alkylidyne Complexes

The first d^0 tantalum alkylidyne complex was prepared by removing the alkylidene α-proton from $Ta(CHCMe_3)(CH_2CMe_3)_3$ with *n*-butyllithium in the presence of a coordinating base (e.g., N, N'-dimethylpiperazine = dmp; Scheme 1) (2). The X-ray crystal structure of $[Li(dmp)][Ta(CCMe_3)(CH_2CMe_3)_3]$

$$Ta(CHCMe_3)(CH_2CMe_3)_3 + LiBu + dmp \xrightarrow{-C_4H_{10}} [Li(dmp)]$$
$$[Ta(CCMe_3)(CH_2CMe_3)_3]$$

Scheme 1.

showed the $Ta \equiv C_\alpha$ triple bond distance to be 176(2) pm and the $Ta\text{-}C_\alpha\text{-}C_\beta$ bond angle to be 165(1)° (Table 1; Figure 1). The lithium atom is within bonding distance of the alkylidyne α-carbon atom (219(3) pm), but on the basis of the coordination geometry about lithium it was felt that the lithium atom was interacting with the electron density in the triple bond, rather than specifically with the alkylidyne carbon atom. In retrospect, this structure foretold the alkylidyne-

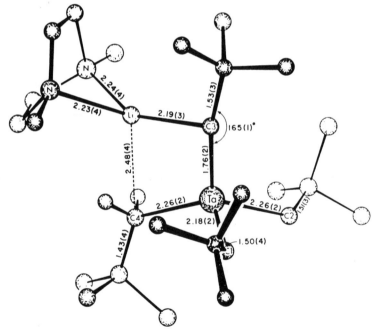

Figure 1. The structure of $[Li(dmp)][Ta(CCMe_3)(CH_2CMe_3)_3]$ (dmp = N,N'-dimethylpiperazine) (2); "Ta≡C(3)" = 176(2) pm.

like character of some grossly distorted neopentylidene complexes of tantalum in which the $Ta-C_\alpha-C_\beta$ angle is large, the $Ta-C_\alpha-H_\alpha$ sometimes less than 90°, and the tantalum-carbon bond approaching the length of a triple bond (3-5). For reference a "normal" tantalum-carbon double bond is about 200 pm (e.g., 203.0(6) pm in $Ta(\eta^5-C_5H_5)_2(CHCMe_3)Cl$ (6) and 203.9(1) pm in $Ta(\eta^5-C_5H_5)_2-(CH_2)(Me))$ (7, 8).

$$Ta(\eta^5-C_5H_5)(CHCMe_3)Cl_2 \xrightarrow[\text{2. } Ph_3P=CH_2]{\text{1. excess } PMe_3} Ta(\eta^5-C_5H_5)(CCMe_3)(PMe_3)_2Cl$$

$$Ta(\eta^5-C_5H_5)(CH_2Ph)_3Cl \xrightarrow[- \text{ 2 PhMe}]{+ \text{ 2 } PMe_3/60\,°C} Ta(\eta^5-C_5H_5)(CPh)(PMe_3)_2Cl$$

$$Ta(\eta^5-C_5Me_5)(CHR)(CH_2R)Cl \xrightarrow[-RCH_3]{2 \text{ } PMe_3} Ta(\eta^5-C_5Me_5)(CR)(PMe_3)_2Cl$$

Scheme 2. $R = CMe_3$ or Ph.

The next tantalum alkylidyne complexes were prepared by removing a proton from an alkylidene ligand in monocyclopentadienyl complexes, as shown in Scheme 2 (9). In the first reaction it was proposed that $Ph_3P = CH_2$ either deprotonates a cationic intermediate, $[Ta(\eta^5-C_5H_5)(CHCMe_3)Cl(PMe_3)_2]^+Cl^-$, or dehydrohalogenates $Ta(\eta^5-C_5H_5)(CHCMe_3)Cl_2(PMe_3)$. The related reaction of $Ta(\eta^5-C_5H_5)(CH_2Ph)_3Cl$ with two equivalents of PMe_3 in benzene at 60 °C gave $Ta(\eta^5-C_5H_5)(CPh)(PMe_3)_2Cl$ and two equivalents of toluene while $Ta(\eta^5-C_5Me_5)(CHCMe_3)(CH_2CMe_3)Cl$ and $Ta(\eta^5-C_5Me_5)(CHPh)(CH_2Ph)Cl$ react immediately with two equivalents of PMe_3 to yield $Ta(\eta^5-C_5Me_5)(CCMe_3)-(PMe_3)_2Cl$ and $Ta(\eta^5-C_5Me_5)(CPh)(PMe_3)_2Cl$, respectively. In the last two reactions it is likely that the alkylidyne ligands are formed via an α-hydrogen abstraction reaction in which the alkyl ligand is employed as the leaving group. It has been observed in other circumstances that PMe_3 accelerates abstraction of an α-hydrogen from the neopentylidene ligand in a crowded intermediate (10). $Ta(\eta^5-C_5H_5)(CCMe_3)(PMe_3)_2Cl$ also was formed when $Ta(\eta^5-C_5Me_5)-(CHCMe_3)(H)(PMe_3)Cl$ (11) was treated with PMe_3 at 60 °C. Here the hydride ligand acts as a leaving group in abstracting the alkylidene α-proton. In $Ta(\eta^5-C_5Me_5)(CPh)(PMe_3)_2Cl$ (Figure 2) (12) the ligands adopt a "four-legged piano stool" arrangement about the metal. The $Ta \equiv C_\alpha$ triple bond distance (184.9(8) pm, Table 1) is ~22 pm, shorter than the $Ta = C_\alpha$ double distance found in $Ta(\eta^5-C_5H_5)_2(CHPh)(CH_2Ph)$ (13). One interesting reaction of these d^0 tantalum alkylidyne complexes, formation of a *bis*(neopentylidene) complex, is shown in Scheme 3 (14).

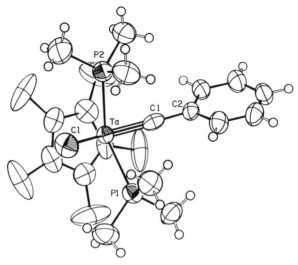

Figure 2. The structure of Ta(η^5-C$_5$Me$_5$)(CPh)(PMe$_3$)$_2$Cl (12); Ta$\equiv$C(1) = 184.9(8) pm.

$$Ta(\eta^5\text{-C}_5R_5)(CCMe_3)(PMe_3)_2Cl \xrightarrow[-2\,PMe_3]{LiCH_2CMe_3} Ta(\eta^5\text{-C}_5R_5)(CHCMe_3)_2Cl$$

Scheme 3. R = H or Me.

Alkylidyne hydride complexes have been prepared by reducing alkylidene complexes. For example, reduction of Ta(η^5-C$_5$Me$_5$)(CHCMe$_3$)Br$_2$ (15–17) with two equivalents of sodium amalgam in the presence of PMe$_3$ gives Ta(η^5-C$_5$Me$_5$)(CCMe$_3$)(H)(PMe$_3$)$_2$ in 80% yield (11). This complex can also be prepared from Ta(η^5-C$_5$Me$_5$)(CH$_2$CMe$_3$)$_2$Cl$_2$, two equivalents of sodium amalgam, and two equivalents of PMe$_3$. Substituting one equivalent of dmpe for the two equivalents of PMe$_3$ in the latter method yields Ta(η^5-C$_5$Me$_5$)(CCMe$_3$)-(H)(dmpe). In one system alkylidyne hydride complexes could be observed in equilibrium with alkylidene complexes (18). An example is shown in Scheme 4; the α-hydrogen atom migrates from a distorted alkylidene ligand ($\nu_{CH\alpha}$ = 2200 cm^{-1}; $J_{CH\alpha}$ = 57 Hz) to give pentagonal bipyramidal Ta(CCMe$_3$)(H)-(dmpe)$_2$(ClAlMe$_3$) (Figure 3). "Ta(CHCMe$_3$)(dmpe)$_2$I" was shown to consist of ~90% Ta(CCMe$_3$)(H)dmpe)$_2$I (in equilibrium with Ta(CHCMe$_3$)(dmpe)$_2$I), while "Ta(CHCMe$_3$)(dmpe)$_2$(O$_3$SCF$_3$)" was shown to consist of at least 50% Ta(CCMe$_3$)(H)(dmpe)$_2$(O$_3$SCF$_3$).

$$Ta(CHCMe_3)(dmpe)_2Cl + AlMe_3) \longrightarrow Ta(CCMe_3)(H)(dmpe)_2$$
$$(ClAlMe_3)$$

Scheme 4.

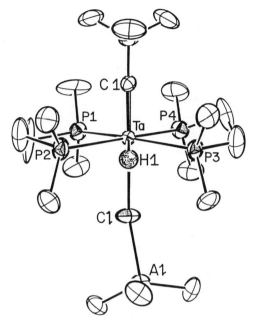

Figure 3. The structure of Ta(CCMe$_3$)(H)(dmpe)$_2$(ClAlMe$_3$) (18); Ta≡C(1) = 185.0(8) pm; Ta-Cl = 275.8(2) pm.

A type of alkylidyne complex that has been known for a relatively long time is [M(CSiMe$_3$)(CH$_2$SiMe$_3$)$_2$]$_2$ (M = Nb or Ta) (19, 20). The niobium compound is the only known niobium alkylidyne complex. This type is included here because it can be viewed as a dimeric version of hypothetical M(CSiMe$_3$)(CH$_2$SiMe$_3$)$_2$, and is related to complexes of the type [Li(L)$_2$][Ta(CCMe$_3$)(CH$_2$CMe$_3$)$_3$] mentioned above. Recently (21) it has been shown that complexes of this type react with phenols, ultimately to give Ta(OAr)$_5$ when OAr = O-2,6-C$_6$H$_3$Me$_2$, but only partially substituted compounds when the phenol is relatively bulky. Ta$_2$(CH$_2$SiMe$_3$)$_x$(OAr)$_{4-x}$(μ-CSiMe$_3$)$_2$ compounds were isolated when OAr = O-2,6-C$_6$H$_3$Ph$_2$ or O-2,6-C$_6$H$_3$(CMe$_3$)$_2$, two of which were structurally characterized.

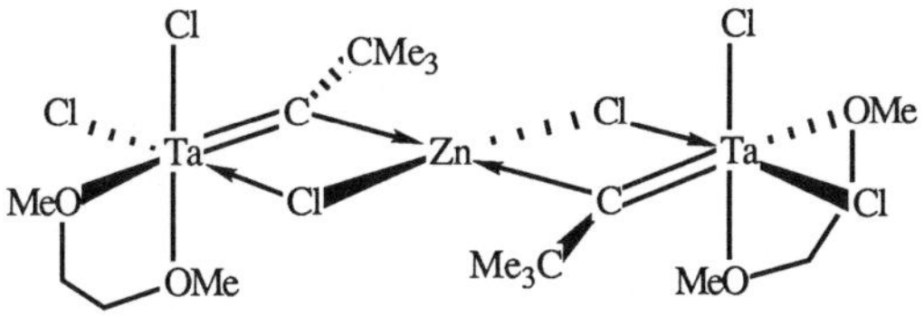

Scheme 5.

Addition of half an equivalent of $Zn(CH_2CMe_3)_2$ to $Ta(CHCMe_3)Cl_3(dme)$ in ether has been shown to yield the trinuclear neopentylidyne complex shown in Scheme 5 (22). The $Ta-C_\alpha$ bond length in $[Ta(\mu-CCMe_3)Cl_2(dme)]_2[Zn(\mu-Cl)_2]$ was found to be about the same as in other tantalum alkylidene complexes, although the $Ta-C_\alpha-C_\beta$ angle is much smaller (Table 1). A plausible intermediate is $[Ta(\mu-CCMe_3)Cl_2(dme)][Zn(\mu-Cl)(CH_2CMe_3)]$, a product of the reaction of $Ta(CHCMe_3)Cl_3(dme)$ with *one* equivalent of $Zn(CH_2CMe_3)_2$ in benzene, since addition of ether to $[Ta(\mu-CCMe_3)Cl_2(dme)][Zn(\mu-Cl)(CH_2CMe_3)]$ yields $[Ta(\mu-CCMe_3)Cl_2(dme)]_2[Zn(\mu-Cl)_2]$. Although $[Ta(\mu-CCMe_3)Cl_2(dme)]_2[Zn-(\mu-Cl)_2]$ does not metathesize internal or terminal alkynes, it reacts with a variety of alkynes to yield cyclopentadienyl complexes (Scheme 6) (23). It is believed that these cyclopentadienyl complexes form via metallacyclobutadiene intermediates $(Ta[C(CMe_3)C(Me)C(Me)]Cl_2(MeC \equiv CMe)$ for the reaction with 2-butyne). In neat 2-butyne a molecule of 2-butyne scavenges $Ta[\eta^5-C_5(CMe_3)-(Me)_4]Cl_2$ to give the Ta(III) alkyne adduct, $Ta[\eta^5-C_5(CMe_3)(Me)_4]Cl_2-(MeC \equiv CMe)$, whereas in dichloromethane $Ta[\eta^5-C_5(CMe_3)(Me)_4]Cl_2$ is oxidized to $Ta[\eta^5-C_5(CMe_3)(Me)_4]Cl_4$. A tantalacyclobutadiene complex, $Ta[C(CMe_3)C(SiMe_3)C(SiMe_3)]Cl_2(Me_3SiC \equiv CSiMe_3)$, has been isolated from the reaction between $[Ta(\mu-CCMe_3)Cl_2(dme)]_2[Zn(\mu-Cl)_2]$ and *bis*(trimethylsilyl)acetylene in dichloromethane (23). Apparently *bis*(trimethylsilyl)-acetylene is too bulky to react with this metallacyle to form a cyclopentadienyl complex such as $Ta[\eta^5-C_5(CMe_3)(SiMe_3)_4]Cl_2(Me_3SiC \equiv CSiMe_3)$ or $Ta[\eta^5-C_5(CMe_3)(SiMe_3)_4]Cl_4$.

$$[Ta(\mu-CCMe_3)Cl_2(dme)]_2[Zn(\mu-Cl)_2] \xrightarrow[\text{25 °C, 80 \%}]{\text{neat 2-butyne}} \begin{array}{l} Ta[\eta^5-C_5(CMe_3) \\ \quad (Me_4)]Cl_2(MeC{\equiv}CMe) \end{array}$$

$$[Ta(\mu-CCMe_3)Cl_2(dme)]_2[Zn(\mu-Cl)_2] \xrightarrow[\text{CH}_2\text{Cl}_2, \text{70 \%}]{\text{2-butyne}} \begin{array}{l} Ta[\eta^5-C_5(CMe_3) \\ \quad (Me_4)]Cl_4 \end{array}$$

Scheme 6.

3   Molybdenum and Tungsten Alkylidyne Complexes

3.1   Alkyl Complexes

W(CCMe$_3$)(CH$_2$CMe$_3$)$_3$ was prepared originally in ~25 % yield by adding six equivalents of LiCH$_2$CMe$_3$ to WCl$_6$ in ether at −78 °C (24). It is now prepared in ~50 % yield by adding W(OMe)$_3$Cl$_3$ to six equivalents of neopentylmagnesium chloride in ether at 0 °C (25). Mo(CCMe$_3$)(CH$_2$CMe$_3$)$_3$ was prepared originally in ~15 % yield by adding five equivalents of LiCH$_2$CMe$_3$ to MoCl$_5$ in ether at −78 °C (24). It is now prepared in 35 ± 5 % yield by adding MoO$_2$Cl$_2$ to six equivalents of neopentylmagnesium chloride in ether at −78 °C (26). Mo-(CCMe$_3$)(CH$_2$CMe$_3$)$_3$ and W(CCMe$_3$)(CH$_2$CMe$_3$)$_3$ are both very soluble in pentane, sensitive to air and moisture, and volatile enough to be sublimed. W(CCMe$_3$)(CH$_2$CMe$_3$)$_3$ can be prepared on a relatively large scale (30–40 g of product), but no more than 7–8 g of Mo(CCMe$_3$)(CH$_2$CMe$_3$)$_3$ can be prepared at a time. Although the mechanism(s) of formation of W(CCMe$_3$)(CH$_2$CMe$_3$)$_3$ and Mo(CCMe$_3$)(CH$_2$CMe$_3$)$_3$ is (are) unknown, in both cases it is believed that a neopentyl ligand is converted into a neopentylidyne ligand by two sequential "α-hydrogen atom abstraction" (3) or α-deprotonation reactions in a M(VI) neopentyl complex (e.g., W(CH$_2$CMe$_3$)$_2$(OMe)$_2$Cl$_2$). Because two controlled α-hydrogen abstraction reactions are involved, and because the neopentyl ligand has been by far the most successful in α-hydrogen abstraction reactions to give tantalum and niobium alkylidene complexes (15–17), this method of preparing trialkyl alkylidyne complexes probably is limited to bulky β-elimination-stabilized alkyls. The only other known trialkyl alkylidyne complexes are Mo(CSiMe$_3$)(CH$_2$SiMe$_3$)$_3$ and W(CSiMe$_3$)(CH$_2$SiMe$_3$)$_3$ (27).

Neither W(CCMe$_3$)(CH$_2$CMe$_3$)$_3$ nor Mo(CCMe$_3$)(CH$_2$CMe$_3$)$_3$ reacts with ordinary alkynes (internal or terminal) in twenty-four hours at room temperature. W(CCMe$_3$)(CH$_2$CMe$_3$)$_3$ does react with phosphines to yield W(CCMe$_3$)(CHCMe$_3$)(CH$_2$CMe$_3$)L$_2$ (L = PMe$_3$ or 0.5 dmpe) and one equivalent of neopentane (Scheme 7) (24). The X-ray crystal structure of W(CCMe$_3$)-(CHCMe$_3$)(CH$_2$CMe$_3$)(dmpe) (Figure 4) (28) shows it to be a distorted square pyramid (with the neopentylidyne ligand in an apical position) in which the

W(CCMe$_3$)(CH$_2$CMe$_3$)$_3$ + 2 L $\longrightarrow$ W(CCMe$_3$)(CHCMe$_3$)
(CH$_2$CMe$_3$)(L)$_2$ + CMe$_4$

Scheme 7. L = PMe$_3$ or 0.5 dmpe.

tungsten-carbon distances are 178.5(8) pm (alkylidyne), 194.2(9) pm (alkylidene), and 225.8(9) pm (alkyl). The corresponding $W-C_\alpha-C_\beta$ bond angles are 175.3(7)°, 150.4(8)°, and 124.5(7)°, respectively. The decreasing tungsten-carbon bond distances and the increasing $W-C_\alpha-C_\beta$ bond angles nicely illustrate the structural differences between a neopentyl, a neopentylidene, and a neopentylidyne ligand, all in the same molecule.

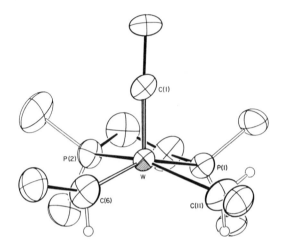

Figure 4. The structure of $W(CCMe_3)(CHCMe_3)(CH_2CMe_3)(dmpe)$ (28); $W\equiv C = 178.5(8)$ pm.

3.2 Halide Complexes

An attempt to remove a phosphine ligand in $W(O)(CHCMe_3)(PEt_3)_2Cl_2$ by oxidizing it with C_2Cl_6 in THF yielded the neopentylidyne complex shown in Scheme 8 (25). In chlorobenzene the reaction proceeds differently. It was proposed that PEt_3Cl_2 actually does form, that it attacks the oxo ligand, ultimately to give $Et_3P = O$, and that an α-proton is removed from the neopentylidene ligand by a phosphine. This discovery demonstrated that neopentylidyne chloride complexes were stable, and therefore led to the more direct methods of synthesis described below.

$$W(O)(CHCMe_3)(PEt_3)_2Cl_2 \xrightarrow[-C_2Cl_4]{C_2Cl_6,\ THF} Et_3PHCl + W(CCMe_3)(Et_3PO)Cl_3$$

$$W(O)(CHCMe_3)(PEt_3)_2Cl_2 \xrightarrow[-C_2Cl_4]{C_2Cl_6,\ PhCl} Et_3PO + [Et_3PH][W(CCMe_3)Cl_4]$$

Scheme 8.

When $M(CCMe_3)(CH_2CMe_3)_3$ (M = Mo (26) or W (25)) is treated with three or more equivalents of anhydrous HCl in the presence of NEt_4Cl, $[Et_4N]$-$[M(CCMe_3)Cl_4]$ is isolated in high yield. If the reaction is carried out in the presence of 1,2-dimethoxyethane (dme), then neutral $M(CCMe_3)Cl_3(dme)$ is isolated in high yield. ^1H and ^{13}C NMR data for $M(CCMe_3)Cl_3(dme)$ at low temperature are consistent with an octahedral structure in which the two ends of the dme ligand are inequivalent. (Dme exchanges readily at 25 °C on the NMR time scale.) It is worth noting that addition of HCl to $M(CCMe_3)(CH_2CMe_3)_3$ (M = Mo or W) results in no net protonation of the metal-carbon triple bond to give either alkylidene or alkyl complexes under mild conditions, even in the presence of excess HCl. The proposed mechanism is shown in Scheme 9. Analogous bromide and iodide derivatives of $M(CCMe_3)Cl_3(dme)$ (M = Mo or W) can be prepared by treating $M(CCMe_3)Cl_3(dme)$ with excess Me_3SiBr (M = Mo (26) or W (29)) or with excess Me_3SiI (M = Mo (29) or W (30)). $Mo(CCMe_3)$-$Br_3(dme)$ can also be prepared from $Mo(CCMe_3)(CH_2CMe_3)_3$ and three equivalents of anhydrous HBr in the presence of dme (26).

$M(CCMe_3)(CH_2CMe_3)_3 + HCl \longrightarrow \text{``}M(CHCMe_3)(CH_2CMe_3)_3Cl\text{''}$

$\text{``}M(CHCMe_3)(CH_2CMe_3)_3Cl\text{''} \longrightarrow CMe_4 + \text{``}M(CCMe_3)(CH_2CMe_3)_2Cl\text{''}$

$\text{``}M(CCMe_3)(CH_2CMe_3)_2Cl\text{''} + 2\ HCl\ etc. \longrightarrow 2\ CMe_4 + \text{``}M(CCMe_3)Cl_3\text{''}$

Scheme 9.

A large number of adducts of neopentylidyne halide complexes have been prepared (see Tables 3 and 4). Some preparations are quite straightforward (e.g., Scheme 10). Many other adducts are compiled in the tables. An X-ray study of $W(CCMe_3)(PMe_3)_3Cl_3$ showed it to be a face-capped octahedron (32). The chemistry of (e.g., phosphine) adducts of "$W(CCMe_3)Cl_3$" (25) is potentially varied and interesting, but is unexplored relative to other areas.

$cis,mer\text{-}Mo(CCMe_3)Cl_3(dme) \xrightarrow{\text{excess py}} cis,mer\text{-}Mo(CCMe_3)Cl_3(py)_2 \ (26)$

$cis,mer\text{-}Mo(CCMe_3)Cl_3(dme) \xrightarrow{2\ PMe_3} cis,fac\text{-}Mo(CCMe_3)Cl_3(PMe_3)_2 \ (31)$

$[Et_4N][W(CCMe_3)Cl_4(PEt_3)] + ZnCl_2 \cdot dioxane \longrightarrow W(CCMe_3)Cl_3(PEt_3) \ (25)$

$W(CCMe_3)Cl_3(dme) + x\ PMe_3 \longrightarrow W(CCMe_3)Cl_3(PMe_3)_x \ (25)$

Scheme 10. x = 1, 2, or 3.

Table 3. Molybdenum Alkylidyne Complexes.

Compound	C_α (ppm)	$^2J_{CP}$ (Hz)	Ref.
$Mo(CCMe_3)(CH_2CMe_3)_3$	323.8		(26)
$Mo(CSiMe_3)(CH_2SiMe_3)_3$	363.9		(27)
$Mo(CCMe_3)Cl_3(dme)$	341.3		(26)
$Mo(CCMe_3)Br_3(dme)$	350.6		(26)
$Mo(CCMe_3)I_3(dme)$	368.6		(29)
$Mo(CCMe_3)Cl_3(PMe_3)_2$	394.5	37	(31)
$Mo(CCMe_3)Cl_3(py)_2$	345.6		(31)
$Mo(CPh)Br_3(dme)$	329.2		(33)
$[Et_4N][Mo(CCMe_3)Cl_4]$	338.8		(26)
$[Et_4N][Mo(CCMe_3)(OCMe_3)Cl_3]$	318.4		(31)
$Mo(CCMe_3)(OCMe_3)Cl_2(PMe_3)$	327.3	24	(31)
$Mo(CCMe_3)(OCMe_3)Cl_2(PEt_3)$	328.6	15	(31)
$Mo(CCMe_3)[OCH(CF_3)_2]_2Cl(dme)$	323.6		(26)
$Mo(CCMe_3)[OCMe(CF_3)_2]_2Cl(dme)$	323.0		(26)
$Mo(CCMe_3)[OC(CF_3)_3]_2Cl(dme)$	335.7		(26)
$Mo(CCMe_3)(OCMe_3)_3$	296.1		(26)
$Mo(CCMe_3)(OCHMe_2)_3$	299.5		(26)
$Mo(CCMe_3)(OCH_2CMe_3)_3$	304.6		(26)
$Mo(CPr)(OCMe_3)_3$	286.6		(26)
$Mo(CCHMe_2)(OCMe_3)_3$	292.7		(26)
$Mo(CPh)(OCMe_3)_3$	276.6		(26)
$Mo(CCMe_3)[OCH(CF_3)_2]_3(dme)$	318.2		(26)
$Mo(CCMe_3)[OCH(CF_3)_2]_3(py)_2$	319.6		(31)
$Mo(CCMe_3)[OCH(CF_3)_2]_3(thf)_2$	318.4		(31)
$Mo(CCMe_3)[OCH(CF_3)_2]_3(thf)$	318.5		(31)
$Mo(CPr)[OCH(CF_3)_2]_3(dme)$	311.6		(26)
$Mo(CPh)[OCH(CF_3)_2]_3(dme)$	297.0		(26)
$Mo(CCMe_3)[OCMe_2(CF_3)]_3$	309.7		(26)
$Mo(CPr)[OCMe_2(CF_3)]_3$	299.3		(26)
$Mo(CCMe_3)[OCMe(CF_3)_2]_3$	325.2		(26)
$Mo(CMe)[OCMe(CF_3)_2]_3$	313.8		(26)
$Mo(CEt)[OCMe(CF_3)_2]_3$	312.1		(26)
$Mo(CPr)[OCMe(CF_3)_2]_3$	313.4		(26)
$Mo(CCMe_3)[OCMe(CF_3)_2]_3(dme)$	318.8		(26)
$Mo(CMe)[OCMe(CF_3)_2]_3(dme)$	302.9		(26)
$Mo(CEt)[OCMe(CF_3)_2]_3(dme)$	310.1		(26)
$Mo(CPr)[OCMe(CF_3)_2]_3(dme)$	309.0		(26)
$Mo(CPh)[OCMe(CF_3)_2]_3(dme)$	295.2		(26)

Compound	C_α (ppm)	$^2J_{CP}$ (Hz)	Ref.
Mo(CCMe₃)[OC(CF₃)₃]₃(dme)			(26)
Mo(CCMe₃)(DIPP)₃	337.2		(26)
Mo(CCMe₃)(DIPP)₃(py)	321.6		(29)
Mo(CCMe₃)(DIPP)₃(PMe₃)	328.1	21	(29)
Mo(CCMe₃)(DIPP)₃(CNCMe₃)	338.2		(29)
Mo(CMe)(DIPP)₃(py)	307.3		(29)
Mo(CEt)(DIPP)₃(py)	314.5		(29)
Mo(CPr)(DIPP)₃(py)	314.5		(29)
Mo(CPh)(DIPP)₃(py)			(29)
Mo(CCMe₃)(O₂CMe)₃	312.4		(75)
Mo(CCMe₃)(O₂CCHMe₂)₃	311.2		(75)
Mo(CCMe₃)(O₂CCHMe₂)₃(PMe₃)	324.9	27	(29)
Mo(CCMe₃)(O₂CCMe₃)₃	311.3		(75)
Mo(CCMe₃)(O₂CCF₃)₃(dme)	341.3		(75)
Mo(CCMe₃)(O₂CCF₃)₃(PMe₃)₂	329.1	20	(29)
Mo(CCMe₃)(NHAr)Cl₂(dme)	321.2		(83)
Mo(CCMe₃)(NHAr)[OCMe(CF₃)₂]₂·			
0.5dme	316.3		(29)
Mo(CCMe₃)(NHAr)(DIPP)₂	317.0		(29)
Mo(CCMe₃)(NMe₂)₃	307.7		(31)
Mo(CCMe₃)(SCMe₃)₃	347.9		(31)
[Mo(CCMe₃)(TMT)₃]₂	336.0		(78)
[Et₃NH][Mo(CCMe₃)(TMT)₃Cl]	316.2		(78)
Mo(CCMe₃)(TMT)₃(PMe₃)	325.4	52	(29)
Mo(CCMe₃)(TMT)₃(py)	330.2		(29)
Mo(CCMe₃)(TMT)₃(CNCMe₃)	328.7		(29)
Mo(CCMe₃)(TMT)₃(PhCHO)	332.1		(29)
Mo(CCMe₃)(TMT)₃(Me₂NCHO)	326.6, 324.6		(29)
Mo(CCMe₃)(TMT)₃(CH₃CN)	330.3		(29)
Mo(CCMe₃)(TIPT)₃	341.5		(78)
[Et₃NH][Mo(CCMe₃)(TIPT)₃Cl]	318.4		(29)
Mo(CCMe₃)(TIPT)₃(PMe₃)	328.5	69	(29)
Mo(CCMe₃)(TIPT)₃(py)	329.5		(29)
Mo(CCMe₃)(TIPT)₃(CNCMe₃)	329.8		(29)
Mo(CCMe₃)(TIPT)₃(PhCHO)	334.4		(29)
Mo(CCMe₃)(TIPT)₃(Me₂NCHO)	326.1		(29)
Mo(CCMe₃)(TIPT)₃(CH₃CN)	330.4		(29)
Mo(CPr)(TIPT)₃	333.6		(78)
Mo(CCMe₃)(PMe₃)₄Cl	284.0	17	(31)

Table 4. Tungsten Alkylidyne Complexes.

Compound	C_α (ppm)	J_{CW} (Hz)	$^2J_{CP}$ (Hz)	Ref.
$W(CCMe_3)(CH_2CMe_3)_3$	316.2	232	21	(25)
$W(CCMe_3)(CHCMe_3)(CH_2CMe_3)$ (dmpe)	296			(24)
$W(CCMe_3)(CHCMe_3)(CH_2CMe_3)$ $(PMe_3)_2$	316	210		(24)
$W(CSiMe_3)(CH_2SiMe_3)_3$	344.6			(27)
$W(CCMe_3)Cl_3(dme)$	335.1	224		(25)
$W(CCMe_3)Br_3(dme)$	346.6			(29)
$W(CCMe_3)I_3(dme)$	365.7			(30)
$W(CMe)Br_3(dme)$	337.5	219		(33)
$W(CCH_2CMe_3)Br_3(dme)$	345.4	215		(33)
$W(CPh)Br_3(dme)$	331.7	219		(33)
$[Et_4N][W(CCMe_3)Cl_4]$	337			(25)
$[Et_4N][W(CCMe_3)Cl_4(PEt_3)]$	335.0		13	(25)
$[Et_4N][W(CCMe_3)Cl_3(NEt_2)]$	297.1	253		(40)
$[Et_4N][W(CCMe_3)Cl_3(N(CHMe_2)_2)]$	295.5	250		(40)
$[Et_3PH][W(CCMe_3)Cl_4]$	339	205		(25)
$W(CCMe_3)Cl_3(PMe_3)$	345		12	(25)
$W(CCMe_3)Cl_3(PMe_3)_2$	357		26	(25)
$W(CCMe_3)Cl_3(PMe_3)_3$	401		40	(25)
$W(CCMe_3)Cl_3(PEt_3)$	346	209	13	(25)
$W(CCMe_3)Cl_3(PPh_3)$	349.0		15	(106)
$W(CCMe_3)Cl_3(Et_3PO)$	329	208		(25)
$W(CCMe_3)Cl_3(PEt_3)(Et_3PO)$	340		15	(25)
$[W(CCMe_3)(OMe)_3]_x$	295.8			(40)
$[W(CCMe_3)(OCHMe_2)_3]_2$	278.6	294		(46)
$W(CCMe_3)(OCMe_3)_3$	271			(25)
$W(CCMe_3)(OCH_2CMe_3)_3$	284.4			(40)
$W(CCMe_3)(OCEt_3)_3$	274.5	294		(40)
$[W(CCMe_3)(OMe)_3(HNMe_2)]_2$	286.0			(25)
$[W(CCMe_3)(OPh)_3(HNMe_2)]_2$	299.7			(40)
$W(CCMe_3)(OCHMe_2)_3(quin)$	272.4			(46)
$[W(CEt)(OCHMe_2)_3]_2$	270.3			(46)
$W(CMe)(OCMe_3)_3$	254.3			(44)
$W(CMe)(OCMe_3)_3(py)$	255.5	298		(44)
$W(CMe)(OCMe_3)_3(quin)$	252.7			(44)
$[W(CMe)(OCMe_3)_3]_2(tmeda)$	253.3			(81)
$W(CEt)(OCMe_3)_3$	262.6			(44)

Compound	C_α (ppm)	J_{CW} (Hz)	$^2J_{CP}$ (Hz)	Ref.
$W(CPr)(OCMe_3)_3$	261.7			(44)
$W(CCHMe_2)(OCMe_3)_3$	268.3			(50)
$W(CCMe_3)(OCMe_3)_3(quin)$	268.7	286		(44)
$W(CCH_2CMe_3)(OCMe_3)_3$	274.5			(50)
$W(CPh)(OCMe_3)_3$	257.0			(44)
$W(CSiMe_3)(OCMe_3)_3$	292.1			(44)
$W(CCH=CH_2)(OCMe_3)_3$	258.0			(44)
$W(CC(Me)=CH_2)(OCMe_3)_3$	260.5			(81)
$W(CCH_2NMe_2)(OCMe_3)_3$	262.7	295		(44)
$W(CCH_2NEt_2)(OCMe_3)_3$	260.1	302		(44)
$W(CCH_2OMe)(OCMe_3)_3(quin)$	253.6	286		(44)
$W(CCH_2OSiMe_3)(OCMe_3)_3(quin)$	254.8	288		(44)
$W[CCH(OEt)_2](OCMe_3)_3(quin)$	254.7	292		(44)
$W(CCO_2Me)(OCMe_3)_3(quin)$	246.8	293		(44)
$W(CCH_2CO_2Me)(OCMe_3)_3(quin)$	240.4	300		(44)
$W[CC(O)Me](OCMe_3)_3(quin)$				(44)
$[W(CNMe_2)(OCMe_3)_3]_2$				(54)
$W(CNEt_2)(OCMe_3)_3(quin)$	223.2	332		(44)
$W(CSMe_3)(OCMe_3)_3$	222.7			(44)
$W(CCH_2CN)(OCMe_3)_3(py)$	235.0	312		(44)
$W(CC\equiv CEt)(OCMe_3)_3(quin)$	232.2	296		(44)
$W(CCN)(OCMe_3)_3(quin)$	215.2	310		(44)
$(Me_3CO)_3W\equiv C-C\equiv W(OCMe_3)_3$	278.6			(44)
$W(CH)(OCMe_3)_3(quin)$	247.1			(44)
$W(CH)(OCMe_3)_3(py)$	252.4			(53)
$W[C-Ru(CO)_2(\eta^5-C_5H_5)](OCMe_3)_3$	237.3	290		(51)
$W(CCMe_3)[OCH(CF_3)_2]_3(dme)$	296.1			(47)
$W(CCMe_3)[OCH(CF_3)_2]_3(py)_2$	299.5			(47)
$W(CPh)[OCH(CF_3)_2]_3(dme)$	281.0			(71)
$W(CEt)[OCH(CF_3)_2]_3(py)_2$	293.6			(47)
$W(CCMe_3)[OCMe_2(CF_3)]_3$	283.0			(46)
$W(CCMe_3)[OCMe_2(CF_3)]_3(py)_2$	284.4			(46)
$W(CMe)[OCMe_2(CF_3)]_3$	265.6			(46)
$W(CMe)]OCMe_2(CF_3)]_3(bipy)$	267.1			(44)
$W(CEt)[OCMe_2(CF_3)]_3$				(46)
$W(CCMe_3)[OCMe(CF_3)_2]_3(dme)$	294.7			(47)
$W(CCMe_3)[OCMe(CF_3)_2]_3(py)_2$	296.0			(47)
$W(CMe)[OCMe(CF_3)_2]_3(dme)$	279.0			(47)
$W(CMe)[OCMe(CF_3)_2]_3(thf)_2$	277.5			(47)

Compound	C_α (ppm)	J_{CW} (Hz)	$^2J_{CP}$ (Hz)	Ref.
W(CEt)[OCMe(CF$_3$)$_2$]$_3$(dme)	286.7	265		(47)
W(CCMe$_3$)(DIPP)$_3$	295.2	289		(45)
W(CCMe$_3$)(DIPP)$_3$(Me$_2$NCHO)	292.4			(95)
W(CCMe$_3$)(DIPP)$_3$(NCCH$_2$CMe$_3$)				(94)
W(CCMe$_3$)(DIPP)$_3$(NCCMe$_3$)	305.7			(94)
W(CCMe$_3$)(DIPP)$_3$(ONC$_5$H$_5$)	300.2			(94)
W(CCMe$_3$)(O-2,6-C$_6$H$_3$Me$_2$)$_3$(thf)	291.4			(45)
W(CMe)(O-2,6-C$_6$H$_3$Me$_2$)$_3$(py)$_2$	284.2			(56)
W(CEt)(O-2,6-C$_6$H$_3$Me$_2$)$_3$(py)$_2$	294.2			(56)
W(CPr)(O-2,6-C$_6$H$_3$Me$_2$)$_3$(py)$_2$	290.5			(56)
W(CPh)(O-2,6-C$_6$H$_3$Me$_2$)$_3$(py)$_2$				(56)
W(CCMe$_3$)(O$_2$CMe)$_3$	286.5			(75)
W(CCMe$_3$)(O$_2$CCHMe$_2$)$_3$	298.2	244		(75)
W(CCMe$_3$)(O$_2$CCMe$_3$)$_3$	285.5			(75)
W(CCMe$_3$)(O$_2$CCF$_3$)$_3$(dme)	319.7			(75)
W(CCMe$_3$)(NHPh)Cl$_2$(PEt$_3$)$_2$	300.5		11	(82)
W(CCMe$_3$)(NHPh)Cl$_2$(PEt$_3$)	304.5		12	(82)
W(CCMe$_3$)(NHPh)Cl$_2$(PMe$_2$Ph)$_2$				(82)
W(CCMe$_3$)(NH$_2$)Cl$_2$(PEt$_3$)$_2$	299		11	(82)
W(CCMe$_3$)(NPh)Cl(PEt$_3$)$_2$	322.7		10	(82)
W(CCMe$_3$)(PHPh)Cl$_2$(PEt$_3$)$_2$	289			(82)
W(CCMe$_3$)(NHAr)Cl$_2$(dme)	304.5			(85)
W(CCMe$_3$)(NHAr)[OCMe(CF$_3$)$_2$]$_2$ (dme)	291.6	279		(85)
W(CCMe$_3$)(NMe$_2$)$_3$	288.3			(25)
W(CEt)(NMe$_2$)$_3$	281.4			(81)
W(CCMe$_3$)[N(SiMe$_3$)$_2$]$_3$				(81)
W(CCMe$_3$)(NMe$_2$)Cl$_2$(PEt$_3$)	313.3		15	(40)
W(CCMe$_3$)(SCMe$_3$)$_3$	334.5			(25)
[W(CCMe$_3$)(TMT)$_3$]$_2$	324.3			(78)
[Et$_3$NH][W(CCMe$_3$)(TMT)$_3$Cl]	304.0	216		(78)
W(CCMe$_3$)(TMT)$_3$(PMe$_3$)	315.5		60	(29)
W(CCMe$_3$)(TMT)$_3$(PhCHO)	322.3			(29)
W(CMe)(TMT)$_3$(PMe$_3$)	304.6		61	(29)
W(CMe)(TMT)$_3$(py)	308.2	224		(78)
W(CCMe$_3$)(TIPT)$_3$	329.4			(78)
[Et$_3$NH][W(CCMe$_3$)(TIPT)$_3$Cl]	306.9			(78)
W(CCMe$_3$)(TIPT)$_3$(PMe$_3$)	318.3		61	(29)

Compound	C_α (ppm)	J_{CW} (Hz)	$^2J_{CP}$ (Hz)	Ref.
$W(CCMe_3)(TIPT)_3(py)$	321.1			(29)
$W(CMe)(TIPT)_3$	312.1			(78)
$[Et_3NH][W(CMe)(TIPT)_3Cl]$	294.8			(78)
$W(CMe)(TIPT)_3(PMe_3)$	308.6		63	(29)
$[Et_3NH][W(CEt)(TIPT)_3Cl]$	301.2			(29)
$W(CEt)(TIPT)_3(PMe_3)$	314.0		64	(29)
$[Et_3NH][W(CPr)(TIPT)_3Cl]$	300.5			(29)
$W(CPr)(TIPT)_3(PMe_3)$				(29)
$W(\eta^5\text{-}C_5H_5)(CCMe_3)Cl_2$	328.3			(88)
$W(\eta^5\text{-}C_5H_5)(CCMe_3)Br_2$	335.8			(31)
$W(\eta^5\text{-}C_5H_5)(CCMe_3)I_2$	346.8			(31)
$W(\eta^5\text{-}C_5H_5)(CCMe_3)(OCMe_3)Cl$	305.7			(31)
$W(\eta^5\text{-}C_5H_5)(CCMe_3)(OMe)Cl$				(31)
$W(\eta^5\text{-}C_5H_5)(CCMe_3)(NEt_2)Cl$	300.7			(88)
$W(\eta^5\text{-}C_5H_5)(CCMe_3)(NHCMe_3)Cl$	301.7			(31)
$W(\eta^5\text{-}C_5H_5)(CCMe_3)[N(SiMe_3)_2]Cl$	306.4			(31)
$W(\eta^5\text{-}C_5H_5)(CCMe_3)(\eta^3\text{-}C_5H_5)Cl$	341.4			(31)
$W(\eta^5\text{-}C_5H_5)(CCMe_3)(OCMe_3)_2$	285.0			(31)
$W(\eta^5\text{-}C_5H_5)(CCMe_3)(OMe)_2$	289.4	290		(31)
$W(\eta^5\text{-}C_5H_5)(CCMe_3)(OCH_2CF_3)_2$	296.4			(31)
$W(\eta^5\text{-}C_5H_5)(CCMe_3)[OCH(CF_3)_2]_2$	302.8			(31)
$W(\eta^5\text{-}C_5H_5)(CCMe_3)(OPh)_2$	298.6	162		(31)
$W(\eta^5\text{-}C_5Me_5)(CPh)(CH_2Ph)_2$	278.8			(93)
$W[\eta^5\text{-}C_5Me_4(CMe_3)](CCMe_3)Cl_2$	317.3			(86)
$W[\eta^5\text{-}C_5Me_4(CMe_3)](CCMe_3)I_2$	335.6			(87)
$\{W[\eta^5\text{-}C_5Me_4(CMe_3)](CCMe_3)Cl\}_2$ $(\mu\text{-}N_2H_2)$	298.7			(87)
$\{W[\eta^5\text{-}C_5Me_4(CMe_3)](CCMe_3)I\}_2$ $(\mu\text{-}N_2H_2)$				(87)
$W(CCMe_3)(PMe_3)_4Cl$	271			(108)
$W(CCMe_3)(dmpe)_2Cl$	271.1		10	(108)
$W(CCMe_3)(H)(dmpe)_2$	281.6	208	10	(108)
$[W(CCMe_3)(H)(dmpe)_2Cl][Cl]$	285.5		11	(108)
$W(CCMe_3)(H)(PMe_3)_3Cl_2$	285	205	15	(111)
$W(CH)(H)(PMe_3)_3Cl_2$	245.7			(108)
$[W(CH)(H)(PMe_3)_3Cl][BH_3CN]$	248.7			(108)
$[W(CH)(H)(dmpe)_2Cl][Cl]$	268.1		12	(108)
$[W(CH)(H)(dmpe)_2Cl][O_3SCF_3]$	287			(108)

Table 5. Rhenium Alkylidyne Complexes.

Compound	C_α (ppm)	Ref.
[Re(CCMe$_3$)(CHCMe$_3$)(NH$_2$CMe$_3$)Cl$_2$]$_2$·thf	294.3, 293.7	(117)
Re(CCMe$_3$)(NHCMe$_3$)(CH$_2$CMe$_3$)(py)Cl$_2$	304.4, 305.0	(117)
[Et$_4$N][Re(CCMe$_3$)(NHCMe$_3$)(CH$_2$CMe$_3$)Cl$_3$]	303.1	(117)
Re(CCMe$_3$)(CHCMe$_3$)(OCMe$_3$)$_2$	287.4	(117)
Re(CCMe$_3$)(CHCMe$_3$)(OSiMe$_3$)$_2$	290.5	(117)
Re(CCMe$_3$)(CHCMe$_3$)(OCMe$_2$CMe$_2$O)(NH$_2$CMe$_3$)	291.9	(117)
Re(CCMe$_3$)(CHCMe$_3$)(CH$_2$CMe$_3$)$_2$	295.1	(117)
Re(CCMe$_3$)(CHCMe$_3$)(CH$_2$SiMe$_3$)$_2$	294.1	(117)
Re(CCMe$_3$)(CHCMe$_3$)(tmeda)I$_2$	292.1	(117)
Re(CCMe$_3$)(CHCMe$_3$)(py)$_2$I$_2$	299.5	(117)
Re(CCMe$_3$)(CH$_2$CMe$_3$)$_3$Cl	278.2	(117)
Re(CCMe$_3$)(CH$_2$CMe$_3$)$_3$(O$_3$SCF$_3$)	296.3	(117)
[H$_3$NAr][Re(CCMe$_3$)(NHAr)Cl$_4$]		(119)
[Et$_4$N][Re(CCMe$_3$)(NHAr)Cl$_4$]	315.9	(119)
[2,4-lutidine·H][Re(CCMe$_3$)(NHAr)Cl$_4$]		(120)
[DBUH][Re(CCMe$_3$)(NHAr)Cl$_4$]	314.7	(120)
Re(CCMe$_3$)(NAr)(OCMe$_3$)$_2$	291.0	(119)
Re(CCMe$_3$)(NAr)[OCMe$_2$(CF$_3$)]$_2$	297.6	(119)
Re(CCMe$_3$)(NAr)[OCMe(CF$_3$)$_2$]$_2$	304.4	(119)
Re(CCMe$_3$)(NAr)[OC(CF$_3$)$_2$CF$_2$CF$_2$CF$_3$]$_2$	315.6	(120)
Re(CCMe$_3$)(NAr)(DIPP)$_2$	304.4	(119)
Re(CCHMe$_2$)(NAr)[OCMe(CF$_3$)$_2$]$_2$	300.5	(119)
Re(CCMe$_3$)(NAr)Cl$_3$(py)		(120)

A potentially usefully route to complexes of the type M(CR)Br$_3$(dme) (M = Mo, R = Ph; M = W, R = Me, Ph, or CH$_2$CMe$_3$) consists of oxidizing M(CR)(CO)$_4$Br with Br$_2$ in dichloromethane in the presence of dme at −78 °C (33). M(CR)(CO)$_4$Br is conveniently prepared by treating [NMe$_4$][(OC)$_5$-MC(O)R] with oxalyl bromide in dichloromethane at −78 °C. This discovery is the first to be published that links low and high oxidation state alkylidyne chemistry via oxidation of low oxidation state species, rather than what is likely to be a more controllable reduction of high oxidation state species.

W(CCMe$_3$)Cl$_3$(dme) reacts with one equivalent of 2-butyne, 3-hexyne, or Me$_3$CC≡CMe to form tungstenacyclobutadiene complexes, W[C(CMe$_3$)-C(R)C(R')]Cl$_3$ (R = R' = Me or Et; R = Me, R' = CMe$_3$) (34, 35). Such species are plausible intermediates in an acetylene metathesis reaction. However, these particular metallacyclobutadiene complexes do not metathesize dialkylalkynes.

Instead, the addition of one or more equivalents of 2-butyne to $W[C(CMe_3)$-$C(Me)C(Me)]Cl_3$ yields a 1:1 mixture of cyclopentadienyl complexes as shown in Scheme 11 (36, 37). $W[\eta^5-C_5(CMe_3)(Me)_4]Cl_2(MeC \equiv CMe)$ has been structurally characterized (38). These reactions are analogous to those noted earlier between $[Ta(\mu-CCMe_3)Cl_2(dme)]_2[Zn(\mu-Cl)_2]$ and internal alkynes (23).

$$W[C(CMe_3)C(Me)C(Me)]Cl_3 \xrightarrow{\ MeC \equiv CMe\ } \text{``} W[\eta^5-C_5(CMe_3)(Me_4)]Cl_3 \text{''}$$

$$\xrightarrow{\ 0.5\,MeC \equiv CMe\ } 0.5\,W[\eta^5-C_5(CMe_3)(Me_4)]Cl_4 + 0.5\,W[\eta^5-C_5(CMe_3)(Me_4)]Cl_2$$
$$(MeC \equiv CMe)$$

Scheme 11.

3.3 Alkoxide and Phenoxide Complexes

The fact that *tert*-butoxide ligands were found to be essential for olefin metathesis activity by one class of niobium and tantalum neopentylidene complexes (39) led to the search for an efficient acetylene metathesis catalyst containing alkoxide ligands. A variety of mono and bis(alkoxide) complexes have been prepared, but in general only as adducts (e. g., PR_3 or pyridine). Many of those listed in the tables have not been published or explored in any detail, since in the absence of basic ligands it is believed that the alkoxide ligands scramble intermolecularly to give mixtures. For this reason, and in order to further discourage bimolecular decomposition reactions and binding of basic ligands (e. g., PR_3) to the metal, tris(alkoxide) complexes were sought.

3.3.1 Preparation from Halide Complexes

A wide variety of $M(CCMe_3)(OR)_3$ species have been prepared by adding three equivalents of LiOR or KOR to $M(CCMe_3)Cl_3(dme)$, or in some cases to $[M(CCMe_3)Cl_4]^-$. The known compounds are listed in the tables, and we will only highlight examples. When the alkoxide is small, e. g., methoxide (40), then only intractable, insoluble materials are obtained; they have been assumed to be polymers formed by alkoxides bridging between metals in pseudo-octahedral coordination environments. At the other extreme compounds containing bulky alkoxides are soluble in hydrocarbons, often pentane. An X-ray structure of $W(CPh)(OCMe_3)_3$ shows it to be a monomer (41) while $W(CMe)(OCMe_3)_3$ contains weakly bridging *tert*-butoxides (42). Depending upon the size and basicity of a base (B) and the size and electron-withdrawing ability of the alkoxide ligand, adducts of the type $M(CCMe_3)(OR)_3(B)$ or $M(CCMe_3)(OR)_3(B)_2$ can be isolated. Subtle differences arise because of finely balanced steric interactions in these crowded molecules. For example, $W(CMe)(OCMe_3)_3(py)$ can be purified by sublimation $(25\,°C, 1.3 \cdot 10^{-3}\,Pa)$ under a *static* vacuum but under a *dynamic* vacuum it loses pyridine to regenerate $W(CMe)(OCMe_3)_3$ (43, 44). $W(CCMe_3)(OCMe_3)_3$ also forms relatively stable $W(CCMe_3)(OCMe_3)_3(quin)$ (quin = 1-azabicyclo[2.2.2]octane) (44), while $W(CCMe_3)(OCMe_3)_3(py)$ readily loses pyridine (a smaller but much weaker donor than quin) *in vacuo*. Many other examples of alkylidyne complexes (e. g., methylidyne or functionalized alkylidyne complexes) that have been isolated only as base adducts may be found in the tables, and some of them in the sections that follow.

Addition of solid $W(CCMe_3)Cl_3(dme)$ to three equivalents of $LiO-2,6-$ $C_6H_3Me_2$ (LiDMP) in tetrahydrofuran yields $W(CCMe_3)(DMP)_3(thf)$ as orange crystals (45). No product has been identified when this reaction is run ether. $W(CCMe_3)(DIPP)_3$, however, can be prepared virtually quantitatively in ether as a yellow-orange waxy solid (Scheme 12). $W(CCMe_3)(DIPP)_3$ does not coordinate THF. It is important to note that complications can arise when the ligand becomes too bulky. For example, $W(CCMe_3)Cl_3(dme)$ reacts with only two equivalents of $LiO-2,6-C_6H_3(CMe_3)_2$ to yield what what is proposed to be an *ortho*-metallated alkylidene complex, $\overline{W(CHCMe_3)[O-2,6-C_6H_3(CMe_2CH_2)-}$ $(CMe_3)][O-2,6-C_6H_3(CMe_3)_2]Cl$, on the basis of ^1H and ^{13}C NMR spectra (45).

$$W(CCMe_3)Cl_3(dme) + 3\ LiDIPP \xrightarrow{\ \text{ether}\ } W(CCMe_3)(DIPP)_3$$

$$W(CCMe_3)Cl_3(dme) + 3\ LiOR_F \longrightarrow W(CCMe_3)(OR_F)_3(dme)$$

Scheme 12. DIPP = $O-2,6-C_6H_3(CHMe_2)_2$; OR_F = $OCH(CF_3)_2$ (47),$OCMe_2(CF_3)$ (46), or $OCMe(CF_3)_2$ (47).

When fluoroalkoxide ligands are added to $W(CCMe_3)Cl_3(dme)$ dimethoxyethane is retained in the coordination sphere (Scheme 12). Sublimation of $W(CCMe_3)[OCMe_2(CF_3)]_3(dme)$ $(60\,°C, 6.6 \cdot 10^{-3}\,Pa)$ results in loss of dme to form $W(CCMe_3)[OCMe_2(CF_3)]_3$ (46). $W(CCMe_3)[OCMe(CF_3)_2]_3(dme)$, on the other hand, sublimes with the dme intact $(60\,°C, 1.3 \cdot 10^{-2}\,Pa)$ (47). It is presumed that $W(CCMe_3)[OCMe_2(CF_3)]_3(dme)$ loses dme *in vacuo* because the metal is not electrophilic enough. Since the $OCMe_2(CF_3)$ ligand is probably significantly smaller than the $OCMe(CF_3)_2$ ligand, the steric properties evidently are more than counterbalanced by the electronic differences. Treating $W(CCMe_3)(OR_F)_3(dme)$ or $W(CCMe_3)[OCMe_2(CF_3)]_3$ with excess pyridine produces the corresponding bis(pyridine) adducts, $W(CCMe_3)(OR_F)_3(py)_2$ [OR_F = $OCH(CF_3)_2$ (47), $OCMe_2(CF_3)$ (46), or $OCMe(CF_3)_2$ (47)]. Pyridine cannot be removed from these compounds *in vacuo* at $25\,°C$.

Syntheses of tris(fluoroalkoxy)neopentylidyne complexes of molybdenum
(26) are not as straightforward as syntheses of the analogous tungsten species.
Reaction conditions (especially solvent and alkali metal ion) must be adjusted
carefully in order to avoid impurities such as the corresponding bis(fluoroalkoxy) complexes. These data suggest that molybdenum is less "electrophilic"
than tungsten, and therefore reacts more slowly with bulky anions that are poor
nucleophiles than does tungsten. Several examples are shown in Scheme 13.

$$Mo(CCMe_3)Cl_3(dme) + 3\ LiOCH(CF_3)_2 \xrightarrow{CH_2Cl_2} Mo(CCMe_3)[OCH(CF_3)_2]_3(dme)$$

$$Mo(CCMe_3)Cl_3(dme) + 3\ KOCMe_2(CF_3) \xrightarrow{ether} Mo(CCMe_3)[OCMe_2(CF_3)]_3$$

$$Mo(CCMe_3)Cl_3(dme) + 3\ KOCMe(CF_3)_2 \xrightarrow{ether} Mo(CCMe_3)[OCMe(CF_3)_2]_3(dme)$$

Scheme 13.

Addition of excess pyridine to $Mo(CCMe_3)[OCH(CF_3)_2]_3(dme)$ yields Mo-
$(CCMe_3)[OCH(CF_3)_2]_3(py)_2$, while dissolving $Mo(CCMe_3)[OCH(CF_3)_2]_3$-
(dme) in tetrahydrofuran followed by removing solvent *in vacuo* yields Mo-
$(CCMe_3)[OCH(CF_3)_2]_3(thf)_2$ (26). $Mo(CCMe_3)[OCH(CF_3)_2]_3(thf)_2$ loses one
THF upon sublimation (60°C and $1.3 \cdot 10^{-3}$ Pa) to yield $Mo(CCMe_3)[OCH$-
$(CF_3)_2]_3(thf)$. The alkoxide ligands in $Mo(CCMe_3)[OCMe_2(CF_3)]_3$ apparently
are not electron withdrawing enough to stabilize $Mo(CCMe_3)[OCMe_2$-
$(CF_3)]_3(dme)$. The complex containing $OCMe(CF_3)_2$ ligands does retain dme,
although dme is lost upon sublimation of $Mo(CCMe_3)[OCMe(CF_3)_2]_3(dme)$
(50°C and $1.3 \cdot 10^{-3}$ Pa) to yield $Mo(CCMe_3)[OCMe(CF_3)_2]_3$. Finally, Mo(CC-
$Me_3)[OC(CF_3)_3]_3(dme)$, which can be prepared from $Mo(CCMe_3)Cl_3(dme)$
and three equivalents of $KOC(CF_3)_3$ in dichloromethane, sublimes with its
coordinated dme intact.

3.3.2 Preparation by Metathetical Reactions Between Alkynes and Alkylidyne Complexes

Many alkoxide alkylidyne complexes react with alkynes to give new alkylidyne complexes, or in some cases (see below) tungstenacyclobutadiene complexes. If the metathesis reaction is fast then the outcome is determined by the
relative volatility of reactants and products, and thermodynamic stability of
various possible alkylidyne complexes.

Tungsten *tert*-butoxide alkylidyne complexes react rapidly with many ordinary alkynes (Scheme 14). Employing an excess of 4-octyne and taking advantage of the greater volatility of $Me_3CC \equiv CPr$ *versus* 4-octyne drives the first reaction completely to $W(CPr)(OCMe_3)_3$ (48, 49). The benzylidyne complex is probably thermodynamically more stable than alkyl-substituted alkylidyne complexes due to conjugation of the phenyl ring with the metal-carbon multiple bond. $W(CC \equiv CEt)(OCMe_3)_3(quin)$ can be prepared from $W(CMe)$-$(OCMe_3)_3$ and $EtC \equiv C-C \equiv CEt$ in the presence of quinuclidine (44). In the absence of base, however, $W(CEt)(OCMe_3)_3$ reacts with $EtC \equiv C-C \equiv CEt$ to yield a dinuclear alkylidyne complex, $(Me_3CO)_3W \equiv C-C \equiv W(OCMe_3)_3$. In the last reaction the coordinated base slows down further metathesis; $W(CC \equiv CEt)(OCMe_3)_3(quin)$ reacts only slowly with $W(CEt)OCMe_3)_3$. Many other examples of alkylidyne complexes prepared by metathetical reactions may be found in the Tables.

$$W(CCMe_3)(OCMe_3)_3 + \text{4-octyne} \longrightarrow W(CPr)(OCMe_3)_3 + Me_3CC \equiv CPr$$

$$W(CCMe_3)(OCMe_3)_3 + PhC \equiv CR \longrightarrow W(CPh)(OCMe_3)_3 \\ + RC \equiv CCMe_3$$

$$W(CMe)(OCMe_3)_3 + Et_2NC \equiv CNEt_2 + quin \xrightarrow{-Et_2NC \equiv CMe} W(CNEt_2) \\ (OCMe_3)_3(quin) \ (44)$$

$$2 \ W(CEt)(OCMe_3)_3 + EtC \equiv C-C \equiv CEt \xrightarrow{-2 \ EtC \equiv CEt} \\ (Me_3CO)_3W \equiv C-C \equiv W(OCMe_3)_3 \ (44)$$

$$W(CMe)(OCMe_3)_3 + EtC \equiv C-C \equiv CEt + quin \xrightarrow{-EtC \equiv CMe} \\ W(CC \equiv CEt)(OCMe_3)_3(quin) \ (44)$$

Scheme 14. R = Ph (48) or Et (49).

$Mo(CCMe_3)(OCMe_3)_3$ reacts only very slowly (hours) with dialkylalkynes, a surprising result in view of the rapid rate at which the tungsten analog reacts (estimated to be of the order of 10^3 min^{-1}; see section 3.5). However, it does react rapidly at $-40\,°C$ with the less sterically demanding terminal alkynes, $RC \equiv CH$ (R = *n*-Pr, Me_2CH, or Ph), as shown in Scheme 15 (26). $Mo(CH)$-$(OCMe_3)_3$ is never formed in these reactions. (Related studies involving tungsten complexes *(vide infra)* suggest that methylidyne complexes are likely to be stable only as base adducts.) As the electron-withdrawing power of the alkoxide increases then molybdenum becomes more reactive toward ordinary dialkylalkynes. For example, $Mo(CCMe_3)[OCMe_2(CF_3)]_3$ reacts with excess

4-octyne to yield $Mo(CPr)[OCMe_2(CF_3)]_3$ and with one equivalent of $PhC\equiv$
CEt to yield $Mo(CPh)[OCMe_2(CF_3)]_3$. Furthermore, $Mo(CCMe_3)[OCMe$-
$(CF_3)_2]_3(dme)$ reacts readily with only slightly more than one equivalent of
either 2-butyne, 3-hexyne, 4-octyne, or $PhC\equiv CEt$ to yield new alkylidyne complexes, $Mo(CR)[OCMe(CF_3)_2]_3(dme)$ (R = Me, Et, *n*-Pr, or Ph, respectively),
while $Mo(CCMe_3)[OCMe(CF_3)_2]_3$ reacts with excess 3-hexyne or 4-octyne in
ether to yield $Mo(CR)[OCMe(CF_3)_2]_3$ (R = Et or *n*-Pr, respectively), or with
2-butyne in toluene to yield $Mo(CMe)[OCMe(CF_3)_2]_3$ (26). Finally, Mo-
$(CCMe_3)[OCH(CF_3)_2]_3(dme)$ reacts with five equivalents of 4-octyne to yield
$Mo(CPr)[OCH(CF_3)_2]_3(dme)$, with one equivalent of $PhC\equiv CEt$ to yield
$Mo(CPh)[OCH(CF_3)_2]_3(dme)$, and with excess (10–20 equivalents) of 2-butyne
or 3-hexyne to yield $Mo(CR)[OCH(CF_3)_2]_3(dme)$ (R = Me or Et, respectively).
The last products are always contaminated with some neopentylidyne complex because the back reaction between $Mo(CR)[OCH(CF_3)_2]_3(dme)$ and
$Me_3CC\equiv CR$ is fast when R = Me relative to the rate of the back reaction when
R = *n*-Pr.

$$Mo(CCMe_3)(OCMe_3)_3 + RC\equiv CH \longrightarrow Mo(CR)(OCMe_3)_3$$
$$+ Me_3CC\equiv CH$$

$$Mo(CCMe_3)(OCMe_3)_3 + RC\equiv CH \longrightarrow X \longrightarrow Mo(CH)(OCMe_3)_3$$
$$+ Me_3CC\equiv CR$$

$$Mo(CCMe_3)[OCMe_2(CF_3)]_3 + PrC\equiv CPr \longrightarrow Mo(CPr)[OCMe_2(CF_3)]_3$$
$$+ Me_3CC\equiv CPr$$

$$Mo(CCMe_3)[OCMe(CF_3)_2]_3(dme) + R'C\equiv CR' \xrightarrow{-Me_3CC\equiv CR'} Mo(CR')$$
$$[OCMe(CF_3)_2]_3(dme)$$

Scheme 15. R = *n*-Pr, $CHMe_2$, or Ph; R' = Me, Et, or *n*-Pr.

Some new alkylidyne complexes can be prepared by the stoichiometric
reaction of alkynes with $W(CCMe_3)(OR_F)_3(dme)$ [$OR_F = OCH(CF_3)_2$ or
$OCMe(CF_3)_2$]. For instance, addition of slightly more than one equivalent of
diphenylacetylene to $W(CCMe_3)[OCH(CF_3)_2]_3(dme)$ yields $W(CPh)[OCH$-
$(CF_3)_2]_3(dme)$ after ~18 hours at room temperature (50). Addition of 2 equivalents of either 2-butyne or 3-hexyne to a pentane solution of $W(CCMe_3)$-
$[OCMe(CF_3)_2]_3(dme)$ yields $W(CR)[OCMe(CF_3)_2]_3(dme)$ (R = Me or Et,
respectively) directly from the solution as yellow crystals at −30 °C (47).
  A novel organoruthenium-substituted methylidyne complex has been prepared by the metathetical reaction shown in Scheme 16 (51). The X-ray crystal

structure of $(Me_3CO)_3W \equiv C\text{-}Ru(CO)_2(\eta^5\text{-}C_5H_5)](OCMe_3)_3$ shows a $W \equiv C_\alpha$-Ru bond angle of 177(2)°, a $W \equiv C_\alpha$ triple bond distance of 175(2) pm, and a Ru-C_α single bond distance of 209(2) pm (Table 1).

$$W(CEt)(OCMe_3)_3 + Ru(C \equiv CMe)(CO)_2(\eta^5\text{-}C_5H_5) \longrightarrow$$
$$(Me_3CO)_3W \equiv C\text{-}Ru(CO)_2(\eta^5\text{-}C_5H_5) + 2\text{-pentyne}$$

Scheme 16.

3.3.3 Preparation by Reactions of Alkynes or Nitriles with Metal-Metal Triple Bonds

An efficient method of preparing many $W(CR)(OCMe_3)_3$ (R ≠ CMe_3) complexes, which avoids $W(CCMe_3)(CH_2CMe_3)_3$ altogether, consists of the metathesis-like reaction of alkynes with the tungsten-tungsten triple bond in $W_2(OCMe_3)_6$ (Scheme 17) (43, 44). For instance, the addition of slightly more than one equivalent of 2-butyne, 3-hexyne, or 4-octyne to $W_2(OCMe_3)_6$ gives $W(CR)(OCMe_3)_3$ (R = Me, Et, or n-Pr) in high yield. $W_2(OCMe_3)_6$ does not react at 25°C with $RC \equiv CR$ (R = CMe_3, $SiMe_3$, $SnMe_3$, or Ph) (44), probably for steric reasons. $W_2(OCMe_3)_6$ does not react readily with diphenylacetylene; at ~70°C polynuclear products (52), as well as small amounts of $W(CPh)$-$(OCMe_3)_3$ (41), are formed. The addition of excess $MeC \equiv CCMe_3$ to $W_2(OCMe_3)_6$ under a static vacuum produces $W(CCMe_3)(OCMe_3)_3$ cleanly because 2-butyne is removed (44). $W(CSiMe_3)(OCMe_3)_3$, free of $W(CMe)$-$(OCMe_3)_3$, cannot be prepared in an analogous manner.

$$W_2(OCMe_3)_6 + RC \equiv CR \longrightarrow 2\ W(CR)(OCMe_3)_3$$

$$W_2(OCMe_3)_6 + 2\ PhC \equiv CEt \longrightarrow 2\ W(CPh)(OCMe_3)_3 + EtC \equiv CEt$$

$$W_2(OCMe_3)_6 + 2\ R'C \equiv CEt \longrightarrow 2\ W(CR')(OCMe_3)_3 + EtC \equiv CEt$$

Scheme 17. R = Me, Et, or n-Pr; R' = CH=CH$_2$ or C(Me)=CH$_2$.

Ethyne (less than 3 equivalents) reacts with $W_2(OCMe_3)_6$ in the presence of pyridine to form $W_2(OCMe_3)_6(\mu\text{-}C_2H_2)(py)$, a dimetallatetrahedrane complex (53). There is good ^{13}C NMR evidence that $W_2(OCMe_3)_6(\mu\text{-}C_2H_2)(py)$ is in equilibrium with the methylidyne complex, $W(CH)(OCMe_3)_3(py)$, in the presence of pyridine. Unfortunately $W(CH)(OCMe_3)_3(py)$ is too unstable to be isolated. A methylidyne complex can be isolated as a stable, but highly oxygen- and moisture-sensitive, quinuclidine adduct from the reaction between $W_2(OCMe_3)_6$ and ethyne in the presence of both pyridine and quinuclidine. The reaction fails in the presence of either base alone (44). These studies suggest that scission of an alkyne can be reversible, and that strongly bound quinuclidine in $W(CR)(OCMe_3)_3(quin)$ complexes may be preventing what is otherwise a favorable recombination reaction.

Metathesis of the tungsten-tungsten triple bond of $W_2(OCMe_3)_6$ provides an opportunity to prepare and study functionalized alkylidyne complexes, some of which are inaccessible using other approaches. Examples are shown in Scheme 18. Base-free analogs of $W(CR")(OCMe_3)_3(quin)$ cannot be prepared cleanly. Moving the carbomethoxy group out of conjugation with the tungsten-carbon triple bond allows one to prepare $W(CCH_2CO_2Me)(OCMe_3)_3$ cleanly from $W_2(OCMe_3)_6$ and two equivalents of $MeC \equiv CCH_2CO_2Me$. Unstable $W[CC(O)Me](OCMe_3)_3(quin)$ can be prepared from $W_2(OCMe_3)_6$ and two equivalents of $EtC \equiv CC(O)Me$ in the presence of both pyridine and quinuclidine. Although $W(CNMe_2)(OCMe_3)_3$ has been prepared from $W_2(OCMe_3)_6$ and Me_2NCN (it is a weakly associated dimer analogous to $[W(CMe)-(OCMe_3)_3]_2$) (54), $W(CNEt_2)(OCMe_3)_3(quin)$ cannot be prepared from $W_2(OCMe_3)_6$ and two equivalents of either $Et_2NC \equiv CNEt_2$ or $MeC \equiv CNEt_2$. In the former case no reaction is observed and in the latter case there is extensive decomposition. Finally, $W_2(OCMe_3)_6$ reacts with two equivalents of $MeC \equiv CSCMe_3$ to yield $W(CSCMe_3)(OCMe_3)_3$ exclusively (44).

$$W_2(OCMe_3)_6 + 1.1\ R_2NCH_2C \equiv CCH_2NR_2 \xrightarrow{\text{pentane}} 2\ W(CCH_2NR_2)(OCMe_3)_3$$

$$W_2(OCMe_3)_6 + 1.1\ R'OCH_2C \equiv CCH_2OR' \xrightarrow{\text{10 quin}} 2\ W(CCH_2OR')(OCMe_3)_3(quin)$$

$$W_2(OCMe_3)_6 + 2\ EtC \equiv CR" + 2\ quin \xrightarrow{-EtC \equiv CEt} 2\ W(CR")(OCMe_3)_3(quin)$$

Scheme 18. R = Me or Et; R' = Me or SiMe$_3$; R" = CH(OEt)$_2$ or CO$_2$Me.

In spite of the relatively large number of compounds containing a metal-metal triple bond (55), few react with alkynes to yield alkylidyne complexes. Two others that do are $W_2(OCHMe_2)_6(py)_2$ and $W_2[OCMe_2(CF_3)]_6$ (46). Addition of one equivalent of 3-hexyne to $W_2(OCHMe_2)_6(py)_2$ gives $[W(CEt)(OCHMe_2)_3]_2$ in good yield after sublimation (25–40 °C, $1.3 \cdot 10^{-3}$ Pa). Alkynes smaller than 3-

hexyne (2-butyne and ethyne), however, react with $W_2(OCHMe_2)_6(py)_2$ to give dimetallatetrahedrane complexes, $W_2(OCHMe_2)_6(\mu\text{-}C_2R_2)(py)_2$ (R = H or Me) (53). $W_2[OCMe_2(CF_3)]_6$ reacts with slightly more than one equivalent of 2-butyne or 3-hexyne to give $W(CMe)[OCMe_2(CF_3)]_3$ (93 % yield after sublimation at 50 °C and $6.6 \cdot 10^{-3}$ Pa) or $W(CEt)[OCMe_2(CF_3)]_3$ (observed by ^1H NMR spectroscopy but not isolated), respectively (46). $W_2(O\text{-}2,6\text{-}C_6H_3Me_2)_6$ also reacts readily with internal alkynes, but the products are tungstenacyclobutadiene complexes (see section 3.4), not alkylidyne complexes, due to the rapid reaction of the incipient alkylidyne complexes with additional alkyne. The addition of excess pyridine to solutions of these metallacyclobutadiene complexes, however, makes it possible to isolate the alkylidyne intermediates as bis(pyridine) adducts, $W(CR)(O\text{-}2,6\text{-}C_6H_3Me_2)_3(py)_2$ (R = Me, Et, *n*-Pr, or Ph) (56).

$Mo_2(OCMe_3)_6$ does not react with disubstituted alkynes to form alkylidyne complexes, but it does react with terminal alkynes to form $Mo(CR)(OCMe_3)_3$ complexes in poor yields. $Mo(CPr)(OCMe_3)_3$ and $Mo(CPh)(OCMe_3)_3$ have been prepared in this manner from 1-pentyne and phenylacetylene, respectively (57). These reactions are not as clean as the analogous reactions of $W_2(OCMe_3)_6$ with internal alkynes because oligomerization and polymerization of 1-pentyne and phenylacetylene compete with the formation of the alkylidyne complexes.

An alternative method of preparing some alkylidyne complexes is to treat $W_2(OCMe_3)_6$ with a nitrile (Scheme 19) (43, 44, 54). In contrast, $W_2[OCMe_2\text{-}(CF_3)]_6$ forms only a labile adduct, $W_2[OCMe_2(CF_3)]_6(CH_3CN)_2$, instead of $W(CMe)[OCMe_2(CF_3)]_3$ and ${W(N)[OCMe_2(CF_3)]_3}_x$ (46). $W_2(DMP)_6$ does not react readily with nitriles either. Reactions between a carbon-nitrogen triple bond and a tungsten-tungsten triple bond, like those between a carbon-carbon triple bond and a tungsten-tungsten triple bond, may fail for a variety of steric, electronic, and structural reasons that are still only poorly understood.

$$W_2(OCMe_3)_6 + RCN \longrightarrow W(CR)(OCMe_3)_3 + 1/x[W(N)(OCMe_3)_3]x$$

Scheme 19. R = a variety of hydrocarbon groups.

3.4 Metallacyclobutadiene Complexes

We noted in section 3.2 that tungstenacyclobutadiene complexes of the type $W[C(CMe_3)C(R)C(R)]Cl_3$ form upon addition of an alkyne to $W(CCMe_3)\text{-}Cl_3(dme)$, and that they react further with alkynes to yield cyclopentadienyl complexes rather that metathesis products. Other tungsten alkylidyne complexes react with alkynes to yield metathesis products (see 3.5), and in some cases tungstenacyclobutadiene complexes can be observed.

The addition of two or more equivalents of 3-hexyne to W(CCMe$_3$)(DIPP)$_3$ yields a tungstenacyclobutadiene complex, W(C$_3$Et$_3$)(DIPP)$_3$ (45). Since α-*tert*-butyl-substituted metallacyclobutadiene complexes are never isolated or observed in this reaction, and since the triethyl-substituted metallacyclobutadiene ring loses 3-hexyne only slowly, steric crowding must force Me$_3$CC≡CEt from W[C(CMe$_3$)C(Et)C(Et)](DIPP)$_3$ to generate unobservable W(CEt)(DIPP)$_3$ which then reacts quickly with the second equivalent of 3-hexyne to yield W(C$_3$Et$_3$)(DIPP)$_3$. W(C$_3$Pr$_3$)(DIPP)$_3$ can be prepared analogously from W(CCMe$_3$)(DIPP)$_3$ and two equivalents of 4-octyne (45). W(C$_3$Et$_3$)(DIPP)$_3$ (Figure 5) has a structure analogous to W[C(CMe$_3$)C(Me)C(Me)]Cl$_3$, a crowded, distorted, trigonal bipyramidal molecule containing a strictly planar, almost symmetrical WC$_3$ ring (34, 35). Alternatively, the molecule can be described as a distorted square pyramid in which the WC$_3$ ring spans the apical and a basal position. The most interesting features of the ring system are the relatively short W-C$_\alpha$ bond lengths, which are roughly halfway between those characteristic of double (195–205 pm) and triple (175–185 pm) tungsten-carbon bonds, and a W···C$_\beta$ distance that is roughly equivalent to a tungsten-carbon single bond length (215–220 pm). The short W···C$_\beta$ distance has been ascribed to the

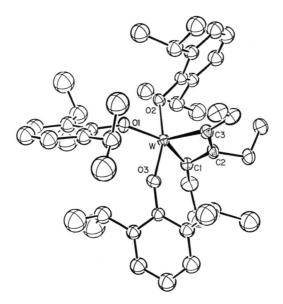

Figure 5. The structure of W(C$_3$Et$_3$)(DIPP)$_3$ (45).

overlap of a tungsten d-orbital with the completely symmetric molecular orbital of the C_3 fragment (58). The tungstenacyclobutadiene ring is slightly distorted, as if 3-hexyne were in the process of leaving the ring.

Tungstenacyclobutadiene complexes also can be isolated from the reactions between $W(CCMe_3)[OCH(CF_3)_2]_3(dme)$ or $W(CCMe_3)[OCMe(CF_3)_2]_3(dme)$ and excess alkyne (47). For instance, when $W(CCMe_3)[OCH(CF_3)_2]_3(dme)$ is treated with 7–10 equivalents of 3-hexyne or 4-octyne at room temperature $W(C_3R_3)[OCH(CF_3)_2]_3$ (R = Et or *n*-Pr, respectively) can be isolated in high yield. Similarly, addition of 20 equivalents of 2-butyne to $W(CCMe_3)[OCMe-(CF_3)_2]_3(dme)$ yields $W(C_3Me_3)[OCMe(CF_3)_2]_3$ as red crystals directly from solution. Addition of bases to these metallacyclobutadiene complexes regenerates alkylidyne complexes (e.g., Scheme 20). The structure of $W(C_3Et_3)-[OCH(CF_3)_2]_3$ (47) is remarkably similar to that of $W(C_3Et_3)(DIPP)_3$ (45). The one significant difference between the two structures is that the WC_3 ring of $W(C_3Et_3)[OCH(CF_3)_2]_3$ shows no structural evidence that 3-hexyne is in the process of being lost. Both are catalysts for the metathesis of alkynes, although by different mechanisms (see section 3.5).

$$W(C_3Et_3)[OCH(CF_3)_2]_3 + \text{excess pyridine} \xrightarrow{\ -EtC{\equiv}CEt\ } \begin{array}{l} W(CEt) \\ [OCH(CF_3)_2]_3(py)_2 \end{array}$$

$$W(C_3Me_3)[OCMe(CF_3)_2]_3 \xrightarrow{\ THF\ } W(CMe)[OCMe(CF_3)_2]_3(thf)$$

Scheme 20.

Molybdenacyclobutadiene complexes, either $Mo(C_3R_3)(OR_F)_3$ or $Mo-[C(CMe_3)C(R)C(R)](OR_F)_3$, have not been observed in reactions of $Mo-(CCMe_3)[OCMe(CF_3)_2]_3(dme)$ with excess alkyne in either ether or pentane (26). So far the only observable molybdenacyclobutadiene complex is $Mo(C_3Et_3)(DIPP)_3$. In solution at −40°C it is in equilibrium with $Mo(CEt)-(DIPP)_3$ and free alkyne. In solution at room temperature it is completely dissociated into $Mo(CEt)(DIPP)_3$ and 3-hexyne.

3.5 Alkyne Metathesis

By analogy with the proposed mechanism for olefin metathesis (59) Katz proposed in 1975 (60) that transition metal carbyne complexes could catalyze the rare acetylene metathesis reaction via unstable metallacyclobutadiene rings (Scheme 21). However, the compounds known at the time to contain a triple bond between a transition metal and carbon, the $M(CR)(CO)_4(X)$ (M = Cr,

Mo, or W; X = halide) complexes first prepared by Fischer (61), did not metathesize alkynes. Since then several alkyne metathesis systems have been discovered, although the nature of the active catalyst in these systems is still
unknown (62–68).

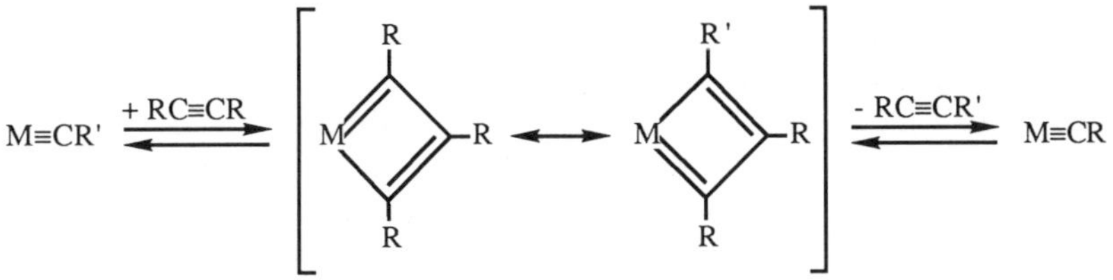

Scheme 21.

W(CCMe₃)(OCMe₃)₃ metathesizes disubstituted alkynes rapidly (48, 49).
The rate of metathesis is fastest (more than ten turnovers per second at 25 °C)
for dialkylalkynes such as 3-heptyne. When the two ends of the alkyne are very
different (either electronically or sterically), the rate of metathesis decreases
dramatically because degenerate metathesis reactions become favored.
Tungstenacyclobutadiene complexes of the type $W(C_3R_3)(OCMe_3)_3$ are never
observed in these metathesis reactions. Addition of $LiOCMe_3$ to the tungstenacyclobutadiene complex, $W[C(CMe_3)C(Me)C(Me)](OCMe_3)Cl_2$ gave a mixture of alkylidyne complexes, $W(CR)(OCMe_3)_3$ ($R = CMe_3$ and Me), as well as
some $W[C(CMe_3)C(Me)C(Me)](OCMe_3)Cl_2$ (36). Similarly, the addition of
the tungstenacyclobutadiene complex, $W(C_3Me_3)Cl_3$ (69), to 3 equivalents
of either $LiOCEt_3$ or $LiOCMe_2Ph$ yields $W(CMe)(OCEt_3)_3$ or $W(CMe)$-
$(OCMe_2Ph)_3$, respectively (70). We conclude that tungstenacyclobutadiene
complexes containing three bulky ordinary alkoxide ligands are unstable with
respect to loss of an alkyne to give an alkylidyne complex.

W(C₃Et₃)(DIPP)₃ will metathesize dialkylalkynes slowly. Since the rate of
incorporation of 3-hexyne-d_{10} into $W(C_3Et_3)(DIPP)_3$ is independent of the concentration of 3-hexyne-d_{10}, the rate-limiting step in any reaction of $W(C_3Et_3)$-
$(DIPP)_3$ with alkynes must be loss of 3-hexyne from $W(C_3Et_3)(DIPP)_3$.

W(C₃Et₃)[OCH(CF₃)₂]₃ metathesizes 20 equivalents of 3-heptyne to equilibrium much faster in ether ($t_{1/2}$ = 10 minutes) than in pentane ($t_{1/2}$ = 21 hours).
The kinetics of the incorporation of 3-hexyne-d_{10} into $W(C_3Et_3)[OCH(CF_3)_2]_3$
are first order in $W(C_3Et_3)[OCH(CF_3)_2]_3$ *and* 3-hexyne-d_{10} in *pentane* (47).
These data are consistent with an *associative* mechanism in which the rate-
limiting step is the formation of a "$W(C_5Et_5)[OCH(CF_3)_2]_3$" intermediate. We
propose that this associative mechanism is favored in the case of $W(C_3Et_3)$-

$[OCH(CF_3)_2]_3$ because the $OCH(CF_3)_2$ ligand is at least as electron-withdrawing as the DIPP ligand but is too small to force the alkyne out of the metallacycle or to prevent coordination of an additional equivalent of an alkyne. In *ether* $W(CEt)[OCH(CF_3)_2]_3(ether)_x$ is postulated to be the most reactive species, although detailed studies as to what extent a dual mechanism might be operative under some conditions have not been carried out.

Any $W(CR)[OCMe(CF_3)_2]_3(dme)$ or $W(C_3R_3)[OCMe(CF_3)_2]_3$ complex metathesizes 20 equivalents of 3-heptyne to equilibrium in less than five minutes in either ether or pentane (47). Although the incorporation of 3-hexyne-d_{10} into $W(C_3Et_3)[OCMe(CF_3)_2]_3$ is too fast to be followed by ^1H NMR spectroscopy, all evidence suggests that $W(C_3Et_3)[OCMe(CF_3)_2]_3$ metathesizes alkynes in non-donor solvents by a *dissociative* mechanism analogous to that involving $W(C_3Et_3)(DIPP)_3$. Apparently, the greater steric bulk of $OCMe(CF_3)_2$ versus $OCH(CF_3)_2$ in $W(C_3Et_3)(OR_F)_3$ is enough to change the mechanism, at least in non-donor solvents, from associative to dissociative.

As mentioned in section 3.3.2, $Mo(CR)(OCMe_3)_3$ (R = *n*-Pr, CMe_3, or Ph) complexes react only very slowly with disubstituted alkynes. $Mo(CCMe_3)$-$(DIPP)_3$, on the other hand, metathesizes 20 equivalents of 3-heptyne to equilibrium in less than five minutes (26). $Mo(CR)(DIPP)_3$ complexes can be isolated as yellow oils from the reactions between $Mo(CCMe_3)(DIPP)_3$ and excess $RC \equiv CR'$ (R = R' = Me, Et, or *n*-Pr; R = Ph, R' = Et) in toluene. Molybdenacyclobutadiene complexes are presumed to be intermediates, although as mentioned earlier, only one, $Mo(C_3Et_3)(DIPP)_3$ (26), has been observed to date.

Molybdenum complexes that contain fluoroalkoxides will metathesize dialkylalkynes at a rate that correlates with the size and electron-withdrawing properties of the alkoxide ligands (26). For example, $Mo(CCMe_3)[OCH(CF_3)_2]_3(dme)$ and $Mo(CCMe_3)[OCH(CF_3)_2]_3(THF)_x$ (x = 1 or 2) metathesize 20 equivalents of 3-heptyne to equilibrium at 25 °C in 5–60 minutes. However, these catalysts also polymerize alkynes, presumably because the $OCH(CF_3)_2$ ligands aren't bulky enough to prevent the intermediate metallacyclobutadiene complexes from reacting with another equivalent of alkyne. The alkoxide ligands in Mo-$(CR)[OCMe_2(CF_3)]_3$ are bulky enough to prevent side reactions such as polymerization but they do not make the metal electrophilic enough to metathesize 3-heptyne very quickly. Hexafluoro-*tert*-butoxide ligands, on the other hand, are bulky enough to prevent polymerization *and* they make the metal in $Mo(CR)[OCMe(CF_3)_2]_3(dme)_x$ (x = 0 or 1) electrophilic enough to metathesize 3-heptyne very rapidly.

Despite the wide range of dialkylacetylene metathesis catalysts available, the metathesis of terminal alkynes has never been observed. One possible reason is that the reaction is likely to be degenerate since the formation of a disubstituted metallacyclobutadiene complex with a proton in the β-position is probably more favorable sterically than one with a proton in the α-position. This seems to be the case with Mo(CCMe₃)(OCMe₃)₃ (Scheme 15). A second reason is that although disubstituted metallacyclobutadiene complexes can be isolated (Scheme 22) (e. g., W[C(CMe₃)C(H)C(CMe₃)][OCH(CF₃)₂]₃ (71) has been structurally characterized (72)), they tend to lose the β-proton to give deprotiometallacyclobutadiene complexes (Scheme 23). Alkylidyne complexes, some of them metathesis catalysts and some of them not, which form deprotiometallacyclobutadiene complexes with terminal alkynes are W(CCMe₃)(OCMe₃)₃ (73, 74), Mo(CCMe₃)(OR$_F$)₃(dme) [OR$_F$ = OCH(CF₃)₂, OCMe(CF₃)₂, or OC(CF₃)₃] (26), W(CR)[OCH(CF₃)₂]₃(dme) (R = Ph or CMe₃) (71), Mo(CR)(DIPP)₃ (R = Et, *n*-Pr, Ph, or CMe₃) (26, 29), W(CCMe₃)(OAr)₃ (OAr = DIPP or O-2,6-C₆H₃Me₂) (71), Mo(CCMe₃)(O₂CCF₃)₃(dme) (75), and W(CCMe₃)-(η⁵-C₅H₅)Cl₂ *(vide infra)* (73). Two of these deprotiometallacyclobutadiene complexes, Mo[C₃(CMe₃)₂][OCH(CF₃)₂]₂(py)₂ (26) and W[C₃(CMe₃)₂](η⁵-C₅H₅)Cl (74), have been structurally characterized. A third possible reason that terminal alkynes are so difficult to metathesize is that methylidyne complexes, which must necessarily form during productive metathesis, are not nearly as stable as substituted methylidyne complexes (44, 53). A fourth reason is that ethyne generated as a product of metathesis is likely to be much more reactive (and in unique ways) than a disubstituted alkyne. Finally, polymerization of terminal alkynes is a significant side reaction (76).

Scheme 22.   OR$_F$ = OCH(CF₃)₂ (71).

Scheme 23.   OR$_F$ = OCH(CF₃)₂ (71).

It has been demonstrated that cyclooctyne can be ring-opened by Mo(CR)-(OCMe$_3$)$_3$ in a metathetical reaction (77). Recent studies show that this polymer is not perfectly living, since Mo(CR)(OCMe$_3$)$_3$ will react with internal alkynes (in this case in the polymer chain) very slowly, although it is relatively well-behaved compared to W(CR)(OCMe$_3$)$_3$, a highly active catalyst for the metathesis of ordinary internal alkynes.

W(CX)(OCMe$_3$)$_3$ complexes in which X is electron-withdrawing (e. g., CO$_2$Me) do not react with 3-heptyne, whereas complexes in which X is σ- or π-electron-donating (e. g. SCMe$_3$ or NEt$_2$) metathesize 3-heptyne readily (44). These data are consistent with a metal-carbon triple bond being polarized M(δ+) ≡ C(δ−) and reaction with an alkyne being interpreted as coordination of an alkyne to an electrophilic metal followed by nucleophilic attack on the alkyne by the alkylidyne α-carbon atom. We shall see below that supporting ligand systems that one would think donate electron density to the metal (e. g. thiolates instead of alkoxides) deactivate the metal toward reaction with an alkyne and are not metathesis catalysts.

These studies suggest that there are at least three mechanisms by which alkynes are metathesized, (i) a dissociative mechanism in which loss of alkyne from the metallacycle is rate limiting, (ii) a solvent-assisted mechanism in which an external base breaks up the metallacycle, and (iii) an associative mechanism in which the metallacycle reacts reversibly with alkyne to yield some as yet unidentified "MC$_5$R$_5$" intermediate. Catalysts that contain alkoxide or phenoxide ligands have been found to be the most active, perhaps in part because a large variety of ligands having various steric and electronic properties are readily available. Alkoxides appear to be able to stabilize monomeric species, yet not deactivate what should be an electrophilic metal by donating too much electron density (cf. amides or thiolates later) or by forming strongly bound dimers. Other factors almost certainly are important, among them the various geometries of intermediates in the catalytic reaction that are accessible, and the ease of intramolecularly rearranging those intermediates.

3.6 Other Supporting Ligand Systems

3.6.1 Carboxylate Complexes

M(CCMe$_3$)(O$_2$CR)$_3$ (M = Mo or W; R = Me, CHMe$_2$, or CMe$_3$) complexes are prepared as shown in Scheme 24 (75). The mechanism of formation of M(CCMe$_3$)(O$_2$CR)$_3$ is thought to be analogous to that for formation of M(CCMe$_3$)Cl$_3$(dme) from M(CCMe$_3$)(CH$_2$CMe$_3$)$_3$ (Scheme 9). The major difference is that RCO$_2$H probably coordinates to the metal via the lone pair of electrons on the carbonyl oxygen atom before protonating the alkylidyne ligand to form "M(CHCMe$_3$)(CH$_2$CMe$_3$)$_3$(O$_2$CR)." No structural studies of M(CCMe$_3$)(O$_2$CR)$_3$ compounds have been completed.

$$M(CCMe_3)(CH_2CMe_3)_3 + 3 RCO_2H \xrightarrow{\text{-3 CMe}_4} M(CCMe_3)(O_2CR)_3$$

Scheme 24. M = Mo or W; R = Me, CHMe$_2$, or CMe$_3$.

M(CCMe$_3$)(O$_2$CR)$_3$ compounds react with disubstituted alkynes to form η^3-cyclopropenyl complexes, M[C$_3$(CMe$_3$)(R')$_2$](O$_2$CR)$_3$ [M = Mo or W; R = Me, CHMe$_2$, or CMe$_3$; R' = Me, Et, *n*-Pr, or Ph], as confirmed by an X-ray structure (75) for W[C$_3$(CMe$_3$)(Et)$_2$](O$_2$CCH$_3$)$_3$. Although these η^3-cyclopropenyl complexes do not metathesize alkynes, they can be induced to lose an alkyne by replacing the carboxylate ligands with alkoxide ligands as shown in Scheme 25. Therefore η^3-cyclopropenyl (or metallatetrahedrane) complexes

$$W[C_3(CMe_3)(Me)_2](O_2CCMe_3)_3 + 3 LiOCMe_3 \longrightarrow W(CR)(OCMe_3)_3$$

Scheme 25. R = CMe$_3$ (50%) and Me (50%).

may be accessible under a variety of conditions, even when only planar metalla-cyclobutadiene rings are observable. At present little is known about the details of the interconversion of a metallatetrahedrane complex and a complex that contains an MC$_3$ ring, and which of the two is closer to the transition state from which an alkyne is lost to give an alkylidyne complex. The issue is complicated by the observation of "bent" metallacyclobutadiene complexes (see later) that could be described as being approximately halfway between the two.

M(CCMe$_3$)(CH$_2$CMe$_3$)$_3$ (M = Mo or W) reacts with three equivalents of trifluoroacetic acid in the presence of dme to form M(CCMe$_3$)(O$_2$CCF$_3$)$_3$(dme) (75). Dimethoxyethane remains bound because the trifluoroacetate ligands are relatively electron withdrawing and likely to be monodentate. Some evidence for monodentate coordination of trifluoroacetate ligands *versus* bidentate coordination of aliphatic carboxylate ligands is the fact that Mo(CCMe$_3$)-(O$_2$CCHMe$_2$)$_3$ reacts with excess PMe$_3$ to form Mo(CCMe$_3$)(O$_2$CCHMe$_2$)$_3$-(PMe$_3$), whereas *cis, mer*-Mo(CCMe$_3$)(O$_2$CCF$_3$)$_3$(dme) reacts with excess PMe$_3$ to form *cis, fac*-Mo(CCMe$_3$)(O$_2$CCF$_3$)$_3$(PMe$_3$)$_2$ (29). None of these tris(trifluoroacetate)neopentylidyne complexes metathesizes alkynes or forms isolable metallatetrahedrane complexes. Instead, W(CCMe$_3$)(O$_2$CCF$_3$)$_3$(dme) reacts with excess 3-hexyne to form a paramagnetic species in low yield which analyzes for W[η^5-C$_5$(CMe$_3$)(Et)$_4$](O$_2$CCF$_3$)$_4$ (75) while Mo(CCMe$_3$)(O$_2$-

$CCF_3)_3(dme)$ reacts with excess 2-butyne to form a diamagnetic species in low yield which appears by ^1H and ^{13}C NMR spectroscopy to be $Mo[\eta^5-C_5(CMe_3)-(Me)_4](O_2CCF_3)_3$ (29).

3.6.2 Thiolate Complexes

The first thiolate alkylidyne complexes to be prepared were $W(CCMe_3)-(SCMe_3)_3$ (25) and $Mo(CCMe_3)(SCMe_3)_3$ (31) (Scheme 26). They do not metathesize 3-heptyne, and the aliphatic thiolate ligands appear to decompose easily via scission of the S-C bond. For this reason, and by analogy with 2,6-diisopropylphenoxide ligands, 2,4,6-trisubstituted arylthiolate ligands were investigated.

$$[Et_4N][W(CCMe_3)Cl_4] + 3 \; LiSCMe_3 \longrightarrow W(CCMe_3)(SCMe_3)_3$$

$$Mo(CCMe_3)Cl_3(dme) + 3 \; HSCMe_3 + 3 \; NEt_3 \longrightarrow Mo(CCMe_3)(SCMe_3)_3$$

Scheme 26.

It is relatively easy to prepare several 2,4,6-trimethylbenzenethiolate (TMT) or 2,4,6-triisopropylbenzenethiolate (TIPT) complexes (Scheme 27) (78). $W(CCMe_3)(SAr)_3$ cannot be prepared from $W(CCMe_3)Cl_3(dme)$ and 3 equivalents of LiSAr, but $W(CCMe_3)(TMT)_3$ can be prepared from $W(CCMe_3)-Cl_3(dme)$ and 3 equivalents of Me_3SiTMT. All $M(CCMe_3)(SAr)_3$ species form mono(adducts) $M(CCMe_3)(SAr)_3(L)$ (29) (M = Mo or W; SAr = TMT or TIPT; L = py, PMe_3, $CNCMe_3$, or Cl^-). All $M(CCMe_3)(SAr)_3(L)$ complexes are proposed to be trigonal bipyramidal species in which the three thiolate ligands occupy equatorial coordination sites (29, 78). Based on variable temperature NMR experiments, the $M(CCMe_3)(TMT)_3$ complexes are dimeric. The $M(CCMe_3)(TIPT)_3$ complexes also may be dimeric, at least in the solid state.

$$M(CCMe_3)Cl_3(dme) + 3 \; ArSH + 3 \; NEt_3 \longrightarrow [Et_3NH][M(CCMe_3)(SAr)_3Cl]$$

$$[Et_3NH][M(CCMe_3)(SAr)_3Cl] + ZnCl_2 \cdot dioxane \longrightarrow M(CCMe_3)(SAr)_3$$

$$Mo(CCMe_3)Cl_3(dme) + 3 \; LiSAr \longrightarrow Mo(CCMe_3)(SAr)_3$$

$$M(CCMe_3)(SAr)_3 + L \longrightarrow M(CCMe_3)(SAr)_3(L)$$

Scheme 27. M = Mo or W; SAr = TMT or TIPT; L = py, PMe_3, or $CNCMe_3$.

M(CCMe$_3$)(SAr)$_3$ complexes do not react with 3-hexyne, 4-octyne, or phenylacetylene, nor do they metathesize 3-heptyne. Only trace amounts of metathesis products are detected after several days at 25 °C (78). Some polymer also forms during these attempted metathesis reactions and the neopentylidyne complexes eventually decompose (probably due to hydrolysis). Tris(thiolate)alkylidyne complexes in which the alkylidyne ligand contains β-protons (Scheme 28) can been prepared, a fact that rules out the instability of such species under metathesis conditions. W(CMe)(TIPT)$_3$ does not metathesize 3-heptyne any better than W(CCMe$_3$)(TIPT)$_3$ so the lack of metathesis activity also cannot be ascribed to the steric bulk of the neopentylidyne ligand.

$$\text{Mo(CPr)[OCMe}_2\text{(CF}_3\text{)]}_3 + 3 \text{ LiTIPT} \longrightarrow \text{Mo(CPr)(TIPT)}_3$$

$$\text{W(C}_3\text{Me}_3\text{)Cl}_3 + 3 \text{ NEt}_3 + 3 \text{ TIPTH} \longrightarrow [\text{Et}_3\text{NH}][\text{W(CMe)(TIPT)}_3\text{Cl}]$$

$$[\text{Et}_3\text{NH}][\text{W(CMe)(TIPT)}_3\text{Cl}] + \text{ZnCl}_2\cdot\text{dioxane} \longrightarrow \text{W(CMe)(TIPT)}_3$$

Scheme 28.

There may be more than one reason why M(CCMe$_3$)(SAr)$_3$ complexes do not metathesize ordinary alkynes. One is that thiolate ligands may donate so much σ-electron density to the metal that the metal is not electrophilic enough to form metallacyclobutadiene complexes. Another is that in hypothetical M(CR)-(SAr)$_3$(alkyne) the alkyne ligand is *trans* to the alkylidyne ligand, while in hypothetical M(CR)(OR)$_3$(alkyne) the acetylene ligand is *cis* to the alkylidyne ligand and thus optimally positioned to react to form a metallacyclobutadiene ring. (These assumptions are based on the X-ray crystal structures of analogous thiolate and phenoxide complexes, Mo(TIPT)$_4$(EtC≡CEt) and Mo(DIPP)$_4$-(PhC≡CPh) (79). Both complexes adopt trigonal bipyramidal geometries but in the structure of the thiolate complex the alkyne ligand occupies an axial position whereas in the structure of the phenoxide complex the alkyne ligand occupies an equatorial position.) Finally, the alkyne in M(CR)(SAr)$_3$(alkyne) may have to compete with a thiolate ligand (in [M(CR)(SAr)$_3$]$_2$) for a coordination site on the metal.

W(C$_3$Me$_3$)(TIPT)$_3$ has been observed as the major product of the reaction between W(C$_3$Me$_3$)Cl$_3$ and three equivalents of LiTIPT (78). Therefore if a metallacyclobutadiene complex were formed in the reaction of M(CCMe$_3$)-(SAr)$_3$ and a dialkylalkyne, it should be stable enough at least to observe spectroscopically. The reactions shown in Scheme 29 also rule out any great kinetic stability of metallacyclobutadiene complexes containing thiolate ligands. Finally, the reaction shown in Scheme 30 suggests that formation of stable

η^3-cyclopropenyl complexes containing thiolate ligands probably is not the reason why thiolate ligands are poor metathesis catalysts.

$$W(C_3Me_3)(TIPT)_3 + PMe_3 \longrightarrow \text{2-butyne} + W(CMe)(TIPT)_3(PMe_3) \quad (29)$$

$$W(C_3R_3)Cl_3 + 3\ NEt_3 + 3\ TIPTH \longrightarrow [Et_3NH][W(CR)(TIPT)_3Cl] \quad (78)$$

Scheme 29. R = Me, Et, or *n*-Pr.

$$W[C_3(CMe_3)(Me)_2]Cl_3(py)_2 + 3\ LiTMT \longrightarrow W(CR)(TMT)_3(py)$$

Scheme 30. R = Me(~60%) and CMe$_3$(~40%).

3.6.3 Amido Complexes

Complexes of the type $M(CCMe_3)(NR_2)_3$ (M = Mo (31) or W (25)) can be prepared from $Mo(CCMe_3)Cl_3(dme)$ or $[Et_4N][W(CCMe_3)Cl_4]$ and three equivalents of $LiNR_2$. $M(CCMe_3)(NMe_2)_3$ complexes do not react with disubstituted alkynes, although $W(CCMe_3)[N(CHMe_2)_2]_3$ has been shown to metathesize phenyltolylacetylene (80). Since $W(CEt)(NMe_2)_3$ that was prepared from $W(CEt)(OCMe_3)_3$ and three equivalents of $LiNMe_2$ (81) did not metathesize 3-heptyne, an impurity such as $W(CCMe_3)[N(CHCMe_2)_2]Cl_2$ was likely to be the actual catalyst in the sample of $W(CCMe_3)[N(CHMe_2)_2]_3$ prepared from $[Et_4N][W(CCMe_3)Cl_4]$.

Amido neopentylidyne complexes can be prepared as shown in Scheme 31 (82). On the basis of ^1H, ^{13}C, and $^{31}P\{^1H\}$ NMR data the amido ligand is believed to be in a position *cis* to the neopentylidyne ligand in order to avoid competition between these two π-bonding ligands for the d-orbitals on the tungsten atom. Heating the amido neopentylidyne complexes results in transfer of the proton on nitrogen to carbon. A five-coordinate amido-alkylidyne analogue, $W(CCMe_3)(NHPh)Cl_2(PEt_3)$, can be prepared from $W(CCMe_3)Cl_3(PEt_3)$ and either LiNHPh or aniline and one equivalent of NEt$_3$, but it decomposes only when heated to 50 °C in benzene for twenty-four hours (82).

The proton transfer reaction in one case (82) has been demonstrated to proceed stepwise as shown in Scheme 32. The α-carbon atom in $W(CCMe_3)$-$(NPh)Cl(PEt_3)_2$ evidently is prone to protonation since the pseudo triple bond of the phenylimido ligand is competing with the tungsten-carbon triple bond.

$[Et_4N][W(CCMe_3)Cl_4] + RNH_2 + NEt_3 + 2\ PEt_3 \longrightarrow$
$$W(CCMe_3)(NHR)Cl_2(PEt_3)_2$$

$[Et_4N][W(CCMe_3)Cl_4] + LiNH_2 + 2\ PEt_3 \longrightarrow W(CCMe_3)(NH_2)Cl_2$
$$(PEt_3)_2$$

$W(CCMe_3)(NHR)Cl_2(PEt_3)_2 \xrightarrow{70\,°C} W(CHCMe_3)(NR)Cl_2(PEt_3)_2$

Scheme 31. R = Ph or H.

$W(CCMe_3)(NHPh)Cl_2(PEt_3)_2 + Ph_3P=CH_2 \longrightarrow W(CCMe_3)(NPh)Cl$
$$(PEt_3)_2 + Ph_3PMeCl$$

$W(CCMe_3)(NPh)Cl(PEt_3)_2 + HCl \longrightarrow W(CHCMe_3)(NPh)Cl_2(PEt_3)_2$

Scheme 32.

$M(CCMe_3)Cl_3(dme) + Me_3SiNHAr \longrightarrow M(CCMe_3)(NHAr)Cl_2$
$$(dme) + Me_3SiCl$$

$M(CCMe_3)(NHAr)Cl_2(dme) \xrightarrow{NEt_3\ cat.} M(CHCMe_3)(NAr)Cl_2(dme)$

Scheme 33. Ar = 2,6-diisopropylphenyl; M = Mo (83) or W (84, 85).

$[Et_4N][W(CCMe_3)Cl_4]$ also reacts with one equivalent of $PhPH_2$ in the presence of two equivalents of PEt_3 to yield $W(CCMe_3)(PHPh)Cl_2(PEt_3)_2$ (82). The NMR spectra of this compound suggest that its structure is analogous to the amido-neopentylidyne complexes. The X-ray crystal structure of $W(CCMe_3)(PHPh)Cl_2(PEt_3)_2$ shows that the phosphido ligand is *cis* to the neopentylidyne ligand and that the phosphine ligands are *trans* to each other. Unlike $W(CCMe_3)(NHR)Cl_2(PEt_3)_2$ (R = Ph or H), however, the phosphido proton in $W(CCMe_3)(PHPh)Cl_2(PEt_3)_2$ does not transfer to the α-carbon atom of the alkylidyne ligand, even after 24 hours at 75 °C.

$M(CCMe_3)Cl_3(dme)$ (M = Mo or W) reacts with one or more equivalents of $Me_3SiNHAr$ as shown in Scheme 33 to yield amido neopentylidyne complexes that can be transformed into imido neopentylidene complexes by catalytic amounts of triethylamine. $M(CCMe_3)(NHAr)Cl_2(dme)$ complexes are valuable intermediates in the preparation of a wide variety of imido neopentylidene complexes (83–85). Transfer of a proton from nitrogen to carbon appears to require the presence of chloride (or presumably other halide) ligands. For example, complexes such as $M(CCMe_3)(NHAr)[OCMe(CF_3)_2]_2(dme)_x$ ($M = Mo$, $x = 0.5$ (29); $M = W$, $x = 1$ (85)) and $Mo(CCMe_3)(NHAr)(DIPP)_2$ (29) have been prepared, but they cannot be transformed into their imido neopentylidene analogs. It is possible that a chloride is required in order to form Et_3NHCl by dehydrohalogenation of the amido ligand. Reprotonation of the neopentylidyne carbon atom would complete the proton transfer. None of the amido-neopentylidyne complexes prepared so far reacts with dialkylalkynes. It is thought that the amido nitrogen donates too much electron density to the metal so that the metal is not electrophilic enough to react with the alkyne.

3.7 Cyclopentadienyl Complexes of Tungsten

Treating $\{W[\eta^5\text{-}C_5(CMe_3)(Me)_4]Cl_4\}_2$ with one equivalent per tungsten of dineopentylzinc yields $W[\eta^5\text{-}C_5(CMe_3)(Me)_4](CCMe_3)Cl_2$ (86). So far it has not been possible to prepare analogous $\eta^5\text{-}C_5Me_5$ complexes from $W(CCMe_3)$-$Cl_3(dme)$ and lithium, sodium, or thallium cyclopentadienyl reagents. An analogous iodide complex, $W[\eta^5\text{-}C_5(CMe_3)(Me)_4](CCMe_3)I_2$, can be made by treating $W[\eta^5\text{-}C_5(CMe_3)(Me)_4](CCMe_3)Cl_2$ with slightly more than two equivalents of Me_3SiI (87). Both of these compounds react with half an equivalent of hydrazine (N_2H_4) in the presence of NEt_3 to yield $\{W[\eta^5\text{-}C_5(CMe_3)(Me)_4]$-$(CCMe_3)(X)\}_2(\mu\text{-}N_2H_2)$ ($X = Cl$ or I), complexes in which the $N_2H_2^{2-}$ ligand bridges both metal centers. The iodide derivative has been structurally characterized (Figure 6) (87). It is interesting to note that conditions have not been found under which the hydrogen atoms of the $\mu\text{-}N_2H_2$ ligand to transfer to the alkylidyne α-carbon atom to yield $\{W[\eta^5\text{-}C_5(CMe_3)(Me)_4](CHCMe_3)(X)\}_2$-$(\mu\text{-}N_2)$.

$W(\eta^5\text{-}C_5H_5(CCMe_3)Cl_2$ can be prepared in high yield in three steps from $[Et_4N][W(CCMe_3)Cl_4]$ as shown in Scheme 34 (88). $W(\eta^5\text{-}C_5H_5)(CCMe_3)Cl_2$ can also be prepared from $W(\eta^5\text{-}C_5H_5)(CCMe_3)(OCMe_3)Cl$ and one equivalent of anhydrous HCl (31). What would appear to be a more straightforward reaction, treating $[Et_4N][W(CCMe_3)Cl_4]$ or $W(CCMe_3)Cl_3(dme)$ with Na, Li, or Tl cyclopentadienyl reagents, led to dark reaction mixtures from which nothing could be isolated. Suitable conditions might be found since the reaction of TlCp ($Cp = \eta^5\text{-}C_5H_5$) with $[Et_4N][W(CCMe_3)Cl_4]$ yielded a mixture of products, one of which could be identified as $W(\eta^5\text{-}C_5H_5)(CCMe_3)Cl_2$ by ^1H NMR spectroscopy. $W(\eta^5\text{-}C_5H_5)(CCMe_3)X_2$ ($X = Br$ or I) can be prepared by treating either

W(η^5-C$_5$H$_5$)(CCMe$_3$)(OCMe$_3$)Cl or W(η^5-C$_5$H$_5$)(CCMe$_3$)Cl$_2$ with excess Me$_3$SiX (X = Br or I) (31).

[Et$_4$N][W(CCMe$_3$)Cl$_4$] + Me$_3$SiNEt$_2$ $\longrightarrow$ [Et$_4$N][W(CCMe$_3$)(NEt$_2$)Cl$_3$]

[Et$_4$N][W(CCMe$_3$)(NEt$_2$)Cl$_3$] + Tl(η^5-C$_5$H$_5$) $\longrightarrow$ W(η^5-C$_5$H$_5$)
                                                                    (CCMe$_3$)Cl(NEt$_2$)

W(η^5-C$_5$H$_5$)(CCMe$_3$)Cl(NEt$_2$) + 2 HCl $\longrightarrow$ W(η^5-C$_5$H$_5$)(CCMe$_3$)Cl$_2$
                                                                    + Et$_2$NH$_2$Cl

Scheme 34.

Addition of excess ROH (OR = OMe, OPh, OCH$_2$CF$_3$, or OCH(CF$_3$)$_2$) to W(η^5-C$_5$H$_5$)(CCMe$_3$)(NEt$_2$)Cl yields W(η^5-C$_5$H$_5$)(CCMe$_3$)(OR)$_2$ (31). Even in the presence of a large excess of ROH there was no evidence for the formation of alkylidene complexes. Although bulky alcohols such as *tert*-butanol do not seem

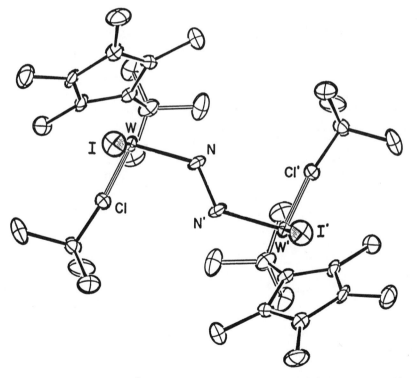

Figure 6. The structure of {W[η^5-C$_5$(CMe$_3$)Me$_4$](CCMe$_3$)I}$_2$(μ-N$_2$H$_2$) (87); W$\equiv$C = 176.9(8) pm.

to react with $W(\eta^5\text{-}C_5H_5)(CCMe_3)(NEt_2)Cl$, $W(\eta^5\text{-}C_5H_5)(CCMe_3)(OCMe_3)_2$ can be prepared from $W(\eta^5\text{-}C_5H_5)(CCMe_3)(OCMe_3)Cl$ and one equivalent of $LiOCMe_3$. $W(\eta^5\text{-}C_5H_5)(CCMe_3)(OMe)Cl$, $W(\eta^5\text{-}C_5H_5)(CCMe_3)(HNCMe_3)Cl$, and $W(\eta^5\text{-}C_5H_5)(CCMe_3)[N(SiMe_3)_2]Cl$ have been prepared in related reactions.

No cyclopentadienyl neopentylidyne complex metathesizes alkynes, but some do react with alkynes. For example, $W(\eta^5\text{-}C_5H_5)(CCMe_3)Cl_2$ reacts with various disubstituted alkynes, $RC \equiv CR'$ ($R = R' = Me$, Et, n-Pr, i-Pr, or Ph; or $R = CMe_3$, $R' = Me$), to form what appear to be unsymmetric, fluxional tungsten-acyclobutadiene complexes (88). The X-ray crystal structure of $W[C(Ph)C-(CMe_3)C(Ph)](\eta^5\text{-}C_5H_5)Cl_2$ (88, 89) shows that the WC_3-ring is nonplanar (the dihedral angle between the C_α-C_β-$C_{\alpha'}$ plane and the C_α-W-$C_{\alpha'}$ plane is ~56°) and that the bonds are localized (alternating single, double, single, double). At this stage it is not known whether all such tungstenacyclobutadiene complexes of this general nature are bent or not. Much work remains to be done in this area, especially in light of recent analogous results in rhenium chemistry *(vide infra)*. It is interesting to note that addition of one equivalent of PMe_3 to $W[C(CMe_3)C-(Me)C(CMe_3)](\eta^5\text{-}C_5H_5)Cl_2$ promotes formation of the η^3-cyclopropenyl complex, $W[C_3(CMe_3)_2(Me)](\eta^5\text{-}C_5H_5)Cl_2(PMe_3)$ (90, 91). One might speculate that the metal is too electrophilic to lose an alkyne, so it retains the C_3 fragment in what is likely to be the sterically smaller η^3-cyclopropenyl form.

$W(\eta^5\text{-}C_5H_5)(CCMe_3)Cl_2$ reacts with one equivalent of 3,3-dimethyl-1-butyne in the presence of one equivalent of NEt_3 to form the deprotiometallacyclobutadiene complex shown in Scheme 35 (73, 74). If NEt_3 is left out of the reaction, $W(\eta^5\text{-}C_5H_5)[C_3(CMe_3)_2]Cl$, and an alkyl(vinyl)-substituted methylene complex, $W(\eta^5\text{-}C_5H_5)[C(CMe_3)CH = CH(CMe_3)]Cl_3$, are formed in equal amounts. $W(\eta^5\text{-}C_5H_5)[C(CMe_3)CH = CH(CMe_3)]Cl_3$ is formed when $W(\eta^5\text{-}C_5H_5)[C_3(CMe_3)_2]Cl$ is treated with two equivalents of anhydrous HCl.

$$W(\eta^5\text{-}C_5H_5)(CCMe_3)Cl_2 + Me_3CC\equiv CH \xrightarrow[\text{-NEt}_3\text{HCl}]{+ \text{ NEt}_3} W(\eta^5\text{-}C_5H_5)[C_3(CMe_3)_2]Cl$$

$$W(\eta^5\text{-}C_5H_5)(CCMe_3)Cl_2 + Me_3CC\equiv CH \longrightarrow 0.5\ W(\eta^5\text{-}C_5H_5)[C_3(CMe_3)_2]Cl + 0.5\ W(\eta^5\text{-}C_5H_5)[C(CMe_3)CH=CH(CMe_3)]Cl_3$$

Scheme 35.

Although $[W(\eta^5\text{-}C_5Me_5)Cl_4]_2$ reacts with four equivalents of $MeMgCl$ to give $W(\eta^5\text{-}C_5Me_5)Me_4$ in high yield (92), it reacts with excess $PhCH_2MgCl$ to yield $W(\eta^5\text{-}C_5Me_5)(CPh)(CH_2Ph)_2$ in moderately high yield (93). The mechanism of forming $W(\eta^5\text{-}C_5Me_5)(CPh)(CH_2Ph)_2$ is not known.

3.8 Other Reactions of Alkylidyne Complexes

3.8.1 Wittig-like Reactions

Some tungsten alkylidyne complexes have been shown to react with nitriles as shown in Scheme 36. This reaction is very sensitive to the steric bulk of the ligands in the alkylidyne complex. For instance, $W(CCMe_3)(OCMe_3)_3$ and some of the functionalized alkylidyne complexes mentioned previously (44) can be recrystallized from ether/acetonitrile mixtures without forming $[W(N)(OCMe_3)_3]_x$. Furthermore, two complexes in which the alkylidyne ligand contains a nitrile functional group have been prepared; $W(CMe)(OCMe_3)_3$ reacts with $EtC \equiv CCH_2CN$ in the presence of pyridine to yield $W(CCH_2CN)$-$(OCMe_3)_3(py)$ while $W(CCN)(OCMe_3)_3(quin)$ can be prepared from $W(CMe)$-$(OCMe_3)_3$ and $EtC \equiv CCN$ (44). Apparently the $W \equiv C$ triple bond reacts more rapidly with the $C \equiv C$ triple bond than it does with the $C \equiv N$ triple bond. Interestingly $W_2(OCMe_3)_6$ reacts with $EtC \equiv CCN$ in the presence of quinuclidine to form $W(CC \equiv CEt)(OCMe_3)_3(quin)$ and $[W(N)(OCMe_3)_3]_x$, rather than $W(CCN)(OCMe_3)_3(quin)$ and $W(CEt)(OCMe_3)_3(quin)$. $W(CCMe_3)$-$(DIPP)_3$ reacts slowly with the more sterically demanding Me_3CCH_2CN (94) while treating $W(CCMe_3)(DIPP)_3$ with the even more sterically demanding Me_3CCN yields only $W(CCMe_3)(DIPP)_3(NCCMe_3)$. Nitride complexes are proposed to form via a planar azametallacyclobutadiene complex in which the nitrogen atom is adjacent to the metal. It is possible that $W(CCMe_3)$-$(DIPP)_3(NCCMe_3)$ is stable because an intermediate azatungstenacyclobu-

$$W(CPr)(OCMe_3)_3 + MeCN \longrightarrow 1/x[W(N)(OCMe_3)_3]_x + PrC \equiv CMe \quad (40)$$

$$W(CCMe_3)(DIPP)_3 + MeCN \longrightarrow 1/x\,[W(N)(DIPP)_3]_x + MeC \equiv CCMe_3 \quad (95)$$

$$W(CMe)(OCMe_3)_3 + EtC \equiv CCN + quin \longrightarrow W(CCN)(OCMe_3)_3(quin) + EtC \equiv CMe \quad (44)$$

$$W_2(OCMe_3)_6 + EtC \equiv CCN \xrightarrow{\ quin\ } 1/x[W(N)(DIPP)_3]_x + W(CC \equiv CEt)(OCMe_3)_3(quin) \quad (44)$$

$$W(CCMe_3)(DIPP)_3 + Me_3CCH_2CN \longrightarrow W(CCMe_3)(DIPP)_3$$
$$(NCCH_2CMe_3) \longrightarrow 1/x[W(N)(DIPP)_3]_x + Me_3CCH_2C \equiv CCMe_3 \quad (94)$$

Scheme 36.

tadiene complex with adjacent *tert*-butyl groups is sterically inaccessible. $Mo(CCMe_3)(SAr)_3$ reacts with acetonitrile to form $Mo(CCMe_3)(SAr)_3$-(CH_3CN) rather than a nitride complex, $[Mo(N)(SAr)_3]_x$ (29).

$W(CCMe_3)(DIPP)_3$ reacts rapidly with one equivalent of acetone to yield an oxo-vinyl complex, $W(O)[C(CMe_3) = CMe_2](DIPP)_3$ (95). This product is believed to form as shown in Scheme 37. Analogous reactions are observed for benzaldehyde, formaldehyde, N,N'-dimethylformamide, and ethyl formate (95). The reaction with dmf is slow enough that $W(CCMe_3)(DIPP)_3(dmf)$ can be isolated and characterized. $W(C_3Et_3)(DIPP)_3$ also reacts with carbonyl compounds in a Wittig-like manner but the reactions are relatively slow since the metallacycle must first lose 3-hexyne to form $W(CEt)(DIPP)_3$. There is some stereoselectivity toward formation of the product in which the largest group (R^1 or R^2) is cis to R (the group in the alkylidyne ligand).

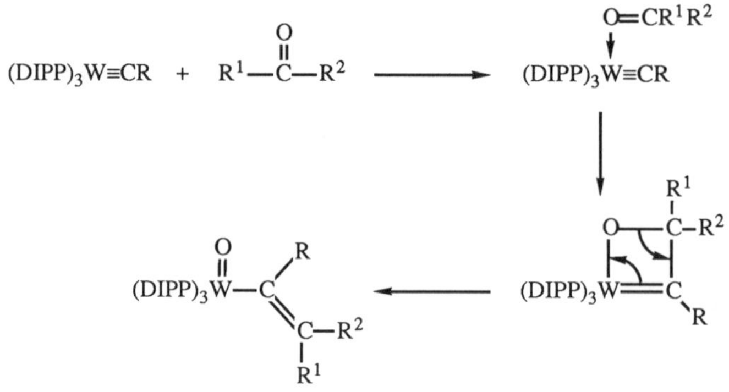

Scheme 37. R = Et or CMe_3; $R^1 = R^2 = $ H or Me; $R^1 = $ H, $R^2 = $ OEt, NMe_2, or Ph.

$W(CCMe_3)Cl_3(dme)$ reacts with two equivalents of cyclohexyl isocyanate to form an oxazetin tungstenacycle, $W[N(Cy)C(C(CMe_3) = C = O)O]Cl_3$ (NCy) (Cy = cyclohexyl) (96). The first step in this reaction is proposed to be the formation of a ketene complex, $W[C(CMe_3) = C = O](NCy)Cl_3$, via an azametallacylobutene intermediate. The second equivalent of cyclohexyl isocyanate is then proposed to insert into the tungsten-ketenyl bond to give the bidentate acylamido ligand in the product.

No molybdenum (VI) alkylidyne complexes to date have been shown to react smoothly in a Wittig-like manner, perhaps because molybdenum is less electrophilic and/or because the Mo $\equiv$ C bond is not as polarized as the W $\equiv$ C bond in the same ligand environment. Both $Mo(CCMe_3)(DIPP)_3$ and $Mo(CCMe_3)$-$(O_2CCF_3)_3(dme)$ react with acetone and benzaldehyde but several (unidentified) products are observed (29). $Mo(CCMe_3)(SAr)_3$ reacts cleanly with one or more equivalents of N, N'-dimethylformamide (dmf) or benzaldehyde to yield mono(adducts), $Mo(CCMe_3)(SAr)_3(L)$. $W(CCMe_3)(TMT)_3$ also reacts with excess benzaldehyde to yield $W(CCMe_3)(TMT)_3(PhCHO)$ (29).

3.8.2 Protonations

Protonation of alkylidyne complexes is still rare, perhaps because incipient alkylidene complexes are susceptible to further protonation, rearrangement, or other forms of decomposition, and because protonation may be reversible. $W(CCMe_3)(OCMe_3)_3$ reacts with two equivalents of HCl, HBr, phenol, or a carboxylic acid to form isolable neopentylidene complexes, $W(CHCMe_3)-(OCMe_3)_2X_2$ (X = Cl, Br, OPh, OC_6F_5, O-p-C_6H_4Cl, O_2CMe, or O_2CPh), and one equivalent of *tert*-butanol (50). The mechanism(s) of the protonation reactions is (are) unknown; at this stage it is assumed that the *tert*-butoxide oxygen atoms are protonated directly. $W(CHCMe_3)(OCMe_3)_2X_2$ complexes metathesize alkenes in the presence of a trace amount of a Lewis acid such as $AlCl_3$; they are analogous to the well-studied class of alkylidene complexes of the type $W(CHCMe_3)(OCH_2CMe_3)_2X_2/AlX_3$ (X = Cl or Br) (97–99).

Complexes in which the alkylidyne ligand contains β-protons also can be protonated to yield alkylidene complexes, though these complexes are much less stable than their neopentylidene analogs. For example, $W(CEt)(OCMe_3)_3$ reacts with two equivalents of phenol or HCl to yield intractable mixtures, but it reacts with two equivalents of pyridinium chloride to yield a stable propylidene complex, $W(CHEt)(OCMe_3)_2Cl_2(py)$ (50). Addition of two equivalents of carboxylic acid (either benzoic acid or acetic acid) to $W(CEt)(OCMe_3)_3$ yields a propylene complex (Scheme 38), $W(CH_2 = CHCH_3)(OCMe_3)_2(O_2CR)_2$, rather than a propylidene complex, $W(CHEt)(OCMe_3)_2(O_2CR)_2$. The alkylidene-to-alkene rearrangement involved in this reaction has been studied in some detail (50). It appears to be catalyzed by excess acid since if only one equivalent of carboxylic acid is added to $W(CEt)(OCMe_3)_3$ then stable propylidene complexes, $W(CHEt)(OCMe_3)_3(O_2CR)$ (R = Me or Ph), are obtained. $W(CHR)-(OCMe_3)_3$(carboxylate) complexes react with alkylidyne complexes of the type $W(CR')(OCMe_3)_3$ to yield $W(CR)(OCMe_3)_3$ and $W(CHR')(OCMe_3)_3$(carboxylate). Thus, alkylidyne complexes can form via deprotonation of alkylidene complexes, *even when the alkylidene ligand contains β-protons.*

$$W(CEt)(OCMe_3)_3 + 2\ RCO_2H \longrightarrow W(CH_2{=}CHCH_3)$$
$$(OCMe_3)_2(O_2CR)_2 + Me_3COH$$

Scheme 38. R = Me or Ph.

Several brief studies have shown that some neopentylidyne complexes react with water relatively cleanly to yield alkyl oxo complexes (Scheme 39). $W(CCMe_3)(CH_2CMe_3)_3$ reacts with excess degassed water to form $[W_2O_3-(CH_2CMe_3)_6]$ (100, 101), a rare d^0 oxo-alkyl complex whose X-ray crystal structure (101) shows it to contain a linear $O = W-O-W = O$ unit in which three neopentyl groups are trigonally disposed about each tungsten atom. Use of D_2O instead of H_2O produces $[W_2O_3(CH_2CMe_3)_4(CD_2CMe_3)_2]$, so the reaction is postulated to occur via double protonation of the neopentylidyne ligand. Other d^0 oxo-alkyl complexes, $[Et_4N][WO_3(CH_2R)]$ (R = CMe_3 or $SiMe_3$), can be prepared by treating $W(CR)(OCMe_3)_3$ with NEt_4OH (102). Analogous compounds cannot be prepared from $W(CR)(OCMe_3)_3$ (R = Et or Ph) and NEt_4OH. Both $[Et_4N][WO_3(CH_2R)]$ (R = CMe_3 or $SiMe_3$) and $[W_2O_3(CH_2-CMe_3)_6]$ are stable in air.

$W(CCMe_3)(CH_2CMe_3)_3$ + excess $H_2O \longrightarrow W_2O_3(CH_2CMe_3)_6$

$W(CR)(OCMe_3)_3$ + excess $NEt_4OH \longrightarrow [Et_4N][WO_3(CH_2R)]$

Scheme 39. R = CMe_3 or $SiMe_3$.

$[Et_4N][W(CCMe_3)Cl_4]$ reacts in tetrahydrofuran with one equivalent of water in the presence of one equivalent of NEt_3 and two equivalents of PEt_3 to form $W(CHCMe_3)(O)(PEt_3)_2Cl_2$, Et_3NHCl, and NEt_4Cl (82). $W(CHCMe_3)(O)-(PEt_3)_2Cl_2$ was originally prepared by a reaction between $W(O)(OCMe_3)_4$ and $Ta(CHCMe_3)Cl_3(PEt_3)_2$ (103–105). The neopentylidyne ligand of $W[\eta^5-C_5(CMe_3)(Me)_4](CCMe_3)Cl_2$ (86) also can be protonated once by water in tetrahydrofuran in the presence of NEt_3 to yield $W[\eta^5-C_5(CMe_3)(Me)_4](CHCMe_3)-(O)Cl$ and Et_3NHCl (106).

3.8.3 Miscellaneous

It has been reported (107) that cyclopentene is polymerized by $W(CCMe_3)-(dme)Cl_3$ while 1-octene is metathesized to ethylene and 7-tetradecene. No organometallic product was isolated, however, so it is still possible that the active catalyst is a small amount of some complex other than $W(CCMe_3)Cl_3(dme)$.

3.9 Some Reactions Involving Lower Oxidation State Alkylidyne Complexes.

Protonation of $W(CCMe_3)(dmpe)_2Cl$ with one equivalent of either HCl or CF_3SO_3H yields cationic neopentylidyne-hydride complexes, $[W(CCMe_3)-(H)(dmpe)_2Cl][X]$ $(X = Cl^-$ or $CF_3SO_3^-)$ (Scheme 40) (108). These complexes react with bases, including $W(CCMe_3)(H)(dmpe)_2$, to give back $W(CCMe_3)-(dmpe)_2Cl$. $W(CCMe_3)(PMe_3)_4Cl$ can also be protonated with triflic acid, but the product is a distorted neopentylidene complex, $[W(CHCMe_3)(PMe_3)_4-Cl][O_3SCF_3]$ (108). This alkylidene complex is analogous to isoelectronic $Ta(CHCMe_3)(PMe_3)_4Cl$ (11) in which the α-hydrogen atom rapidly migrates around the four CP_2 faces in one half of the molecule. The differences between $[W(CHCMe_3)(PMe_3)_4Cl][O_3SCF_3]$ and $[W(CCMe_3)(H)(dmpe)_2Cl][O_3SCF_3]$ can be ascribed to steric hindrance; the four PMe_3 ligands cannot lie in a pentagonal plane along with the hydride ligand in hypothetical pentagonal-bipyramidal $[W(CCMe_3)(H)(PMe_3)_4Cl][O_3SCF_3]$, whereas two more compact dmpe ligands can. Even four equatorial PMe_3 ligands without the added congestion of a hydride ligand cannot lie flat. For example, the PMe_3 ligands in $W(CH)-(PMe_3)_4Cl$ *(vide infra)* (109) and $W(CMe)(PMe_3)_4(Me)$ (110) must pucker in and out of the WP_4-plane in order to avoid adverse steric interactions.

$$W(CCMe_3)(dmpe)_2Cl + HX \longrightarrow [W(CCMe_3)(H)(dmpe)_2Cl][X]$$

$$W(CCMe_3)(PMe_3)_4Cl + CF_3SO_3H \longrightarrow [W(CHCMe_3)(PMe_3)_4Cl]$$
$$[O_3SCF_3]$$

Scheme 40. $X = Cl$ or CF_3SO_3.

An unusual reaction is that between $W(CCMe_3)Cl_3(PMe_3)_3$ and molecular hydrogen (207 kPa for 12 hours) (111) (Scheme 41). $W(CHCMe_3)(H)Cl_3(PMe_3)_2$ reacts with carbon monoxide to yield $W(CHCMe_3)(CO)Cl_3(PMe_3)_2$ and one equivalent of HCl, and with two equivalents of PMe_3 to yield $W(CCMe_3)(H)-(PMe_3)_3Cl_2$ and Me_3PHCl (Scheme 41). $W(CCMe_3)(H)(PMe_3)_3Cl_2$, in turn, reacts with either carbon monoxide or ethylene to yield $W(CHCMe_3)(X)Cl_2-(PMe_3)_2$ $(X = CO$ or $C_2H_4)$ and one equivalent of PMe_3. On the basis of these as yet poorly explored reactions one might conclude that alkylidene complexes are favored in the presence of good π-acceptor ligands, whereas alkylidyne-hydride complexes are favored in their absence.

$$W(CCMe_3)Cl_3(PMe_3)_3 + H_2 \longrightarrow W(CHCMe_3)(H)Cl_3(PMe_3)_2 + PMe_3$$

$$W(CHCMe_3)(H)Cl_3(PMe_3)_2 + 2\ PMe_3 \longrightarrow W(CCMe_3)(H)Cl_2(PMe_3)_3$$
$$+ Me_3PHCl$$

Scheme 41.

Both $W(CH)(PMe_3)_4Cl$ and $W(CH)(dmpe)_2Cl$ can be protonated with triflic acid (Scheme 42) (108, 112). $[W(CH)(H)(dmpe)_2Cl][O_3SCF_3]$ is analogous to $[W(CCMe_3)(H)(dmpe)_2Cl][O_3SCF_3]$ *(vide supra)* (108). $[W(CH_2)(PMe_3)_4$-$Cl][O_3SCF_3]$ is best described as a distorted methylene complex or a face-protonated methylidyne complex. The same steric factors that distinguished $[W(CHCMe_3)(PMe_3)_4Cl][O_3SCF_3]$ from $[W(CCMe_3)(H)(dmpe)_2Cl][O_3SCF_3]$ *(vide supra)* also distinguish $[W(CH_2)(PMe_3)_4Cl][O_3SCF_3]$ from $[W(CH)(H)$-$(dmpe)_2Cl][O_3SCF_3]$. Both $W(CH)(PMe_3)_4Cl$ and $W(CH)(dmpe)_2Cl$ also can be protonated with one equivalent of HCl (Scheme 42). Whereas $W(CH)$-$(dmpe)_2Cl$ yields a methylidyne-hydride complex, $[W(CH)(H)(dmpe)_2Cl][Cl]$, analogous to the product of the triflic acid reaction, $W(CH)(PMe_3)_4Cl$ also yields a methylidyne-hydride complex, $W(CH)(H)(PMe_3)_3Cl_2$, made possible by loss of a phosphine.

$$W(CH)(dmpe)_2Cl + HX \longrightarrow [W(CH)(H)(dmpe)_2Cl][X]$$

$$W(CH)(PMe_3)_4Cl + CF_3SO_3H \longrightarrow [W(CH_2)(PMe_3)_4Cl][O_3SCF_3]$$

$$W(CH)(PMe_3)_4Cl + HCl \longrightarrow W(CH)(H)(PMe_3)_3Cl_2 + PMe_3$$

Scheme 42. $X = Cl$ or CF_3SO_3.

$W(CH)(PMe_3)_4X$ ($X = I$ or O_3SCF_3) complexes can also be protonated with CF_3SO_3H to yield $[W(CH_2)(PMe_3)_4X][O_3SCF_3]$ (113), compounds that are analogous by NMR and IR spectroscopy to $[W(CH_2)(PMe_3)_4Cl][O_3SCF_3]$. The iodide derivative has a trans-octahedral structure (114). In a few cases these distorted methylene complexes can be deprotonated to reform methylidyne complexes. For example, $[W(CH_2)(PMe_3)_4(O_3SCF_3)][O_3SCF_3]$ reacts with $NaBH_4$ to yield $W(CH)(PMe_3)_4(O_3SCF_3)$, and with ethylene to yield what appears by NMR spectroscopy to be $W(CH)(PMe_3)_3(C_2H_4)(O_3SCF_3)$ (113). $[W(CH_2)(PMe_3)_4Cl][O_3SCF_3]$ reacts with one equivalent of $NaBH_3CN$ to yield $[W(CH)(H)(PMe_3)_3Cl][BH_3CN]$ (108).

$W(CH)(H)(PMe_3)_3Cl_2$ reacts with carbon monoxide to yield a methylene-phosphorane complex (Scheme 43) (113). This reaction is proposed to proceed by way of an intermediate methylene complex ("$W(CH_2)(PMe_3)_3(CO)Cl_2$")

because $[W(CH_2)(PMe_3)_4Cl][O_3SCF_3]$ reacts with CO to yield $[W(CH_2PMe_3)-(PMe_3)_3(CO)_2Cl][O_3SCF_3]$ (115). The formation of $W(CH_2PMe_3)(PMe_3)_2-(CO)_2Cl_2$ from $W(CH)(H)(PMe_3)_3Cl_2$ and excess CO resembles the formation of $W(CHCMe_3)(PMe_3)_2(CO)Cl_2$ from $W(CCMe_3)(H)(PMe_3)_3Cl_2$ and excess CO to the extent that the reactive tautomer is likely to be the alkylidene form rather than the alkylidyne-hydride form. The reason why reaction between $W(CH)(H)(PMe_3)_3Cl_2$ and CO doesn't stop at $W(CH_2)(PMe_3)_2(CO)Cl_2$ is not clear.

$$W(CH)(H)(PMe_3)_3Cl_2 + CO \longrightarrow \text{``}W(CH_2)(PMe_3)_3(CO)Cl_2\text{''}$$

$$\text{``}W(CH_2)(PMe_3)_3(CO)Cl_2\text{''} + CO \longrightarrow W(CH_2PMe_3)(PMe_3)_2(CO)_2Cl_2 + PMe_3$$

Scheme 43.

Finally, in an attempt to prepare a tungsten (VI) compound such as $W(CH)Cl_3(PMe_3)_3$ from $W(CH)(PMe_3)_4Cl$, $W(CH)(PMe_3)_4Cl$ was treated with hexachloroethane in the presence of $AlCl_3$ to give a low yield of a phosphino-methylene complex, $[W(CPMe_3)(PMe_3)_2Cl_2]_2[AlCl_4]_2$ (113). This compound is probably most accurately viewed as a W(V) complex; it has been structurally characterized (113).

4 Rhenium Complexes

Since d^0 alkylidyne complexes of tungsten and molybdenum seemed to be relatively common and stable (under the right circumstances), it seemed likely that rhenium alkylidyne complexes in their highest possible oxidation state might be prepared. One might expect Re(VII) to be reduced more easily than W(VI) or Mo(VI), so gaining entry into an organometallic Re(VII) manifold becomes problematic. Replacing oxo ligands by imido ligands before the metal–carbon bond is formed has been found to be a successful method. Perhaps the more electron-donating imido ligands stabilize Re(VII) toward reduction. Imido ligands are also likely to be less prone to attack for steric reasons.

Re(CHCMe$_3$)(CH$_2$CMe$_3$)(NCMe$_3$)$_2$ reacts with three equivalents of 2,4-lutidine·HCl to form [Re(CCMe$_3$)(CHCMe$_3$)(NH$_2$CMe$_3$)Cl$_2$]$_2$, a dimer held together by bridging chloride ligands (116, 117). The neopentylidyne and neopentylidene ligands are mutually cis in order to minimize competition for the d-orbitals on the rhenium. Re(CHCMe$_3$)(CH$_2$CMe$_3$)(NCMe$_3$)$_2$ also reacts with three equivalents of pyridine·HCl to form thermally unstable Re(CCMe$_3$)-(CH$_2$CMe$_3$)(NHCMe$_3$)Cl$_2$(py) (116, 117) in which the amido ligand and the alkylidyne ligand are mutually cis. This type of reaction is interesting because there is overall an apparent transfer of one of the α-protons of the neopentyl ligand *to* the amido nitrogen atom, the opposite of what is observed for W(CCMe$_3$)(NHPh)(PMe$_3$)$_2$Cl$_2$ (82) and M(CCMe$_3$)(NHAr)Cl$_2$(dme) (M = Mo (83) or W (84, 85)). Some related reactions are shown in Scheme 44.

Re(CCMe$_3$)(CH$_2$CMe$_3$)(NHCMe$_3$)Cl$_2$(py) + 2,4-lutidine·HCl + NEt$_4$Cl
$$\longrightarrow \text{[Et}_4\text{N][Re(CCMe}_3\text{)(CH}_2\text{CMe}_3\text{)(NHCMe}_3\text{)Cl}_3\text{]}$$

[Re(CCMe$_3$)(CHCMe$_3$)(NH$_2$CMe$_3$)Cl$_2$]$_2$ + 2 LiX $\longrightarrow$
$$\text{Re(CCMe}_3\text{)(CHCMe}_3\text{)X}_2$$

Re(CCMe$_3$)(CHCMe$_3$)(CH$_2$CMe$_3$)$_2$ + HY $\longrightarrow$ Re(CCMe$_3$)
$$\text{(CH}_2\text{CMe}_3\text{)}_3\text{Y}$$

Re(CCMe$_3$)(CH$_2$CMe$_3$)$_3$Cl + Me$_3$SiO$_3$SCF$_3$ $\longrightarrow$ Re(CCMe$_3$)
$$\text{(CH}_2\text{CMe}_3\text{)}_3\text{(O}_3\text{SCF}_3\text{)}$$

Re(CCMe$_3$)(CHCMe$_3$)(OCMe$_3$)$_2$ + 2 Me$_3$SiI + 2 py $\longrightarrow$ Re(CCMe$_3$)
$$\text{(CHCMe}_3\text{)I}_2\text{(py)}_2$$

Scheme 44. X = OCMe$_3$, OSiMe$_3$, CH$_2$CMe$_3$, or CH$_2$SiMe$_3$; Y = Cl or I.

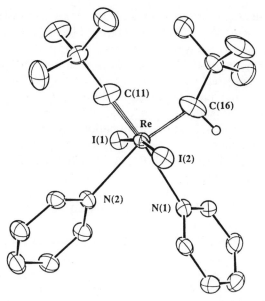

Figure 7. The structure of Re(CCMe₃)(CHCMe₃)I₂(py)₂ (117); Re≡C(11) = 174.2(9) pm; Re=C(16) = 187.3(9) pm; Re=C(16)-C(17) = 150.3(7)°.

The structure of Re(CCMe₃)(CHCMe₃)I₂(py)₂ (117)' is shown in Figure 7. The rhenium-carbon triple bond distance, 174.2(9) pm (Table 1), is significantly shorter than the rhenium-carbon double bond distance, 187.3(9) pm. Furthermore, the rhenium-carbon triple bond distance is shorter than any of the tantalum-carbon or tungsten-carbon triple bond distances listed in Table 1. This result probably can be ascribed to the smaller covalent radius of rhenium(VII) compared to tantalum(V) or tungsten(VI). The Re-N bond trans to the alkylidyne ligand (241(6) pm) is 5 pm longer than the Re-N bond trans to the alkylidene ligand (236.9(6) pm), as has been oberved in other circumstances. For example, in W(CCMe₃)(CHCMe₃)(CH₂CMe₃)(dmpe) (28) the W-P bond distance trans to the neopentylidene ligand (257.7(3) pm) is longer than the W-P bond distance trans to the neopentyl ligand (245.0(3) pm), and in W(CCMe₃)-(PHPh)(PMe₃)₂Cl₂ (82) the W-Cl bond distance trans to the neopentylidyne ligands (257.8(2) pm) is longer than the W-Cl bond distance trans to the phosphido ligand (245.6(2) pm). None of these rhenium (VII) alkylidene or alkylidyne complexes metathesizes alkenes or alkynes, possibly because the metal is not electrophilic enough.

The success of the molybdenum (83) and tungsten (84, 85) olefin metathesis catalyst containing 2,6-diisopropylphenylimido ligands (NAr) prompted an attempt to prepare analogous complexes of Re(VII). This effort has recently been successful (Scheme 45) (118). The reaction in which $[ArNH_3][Re(CCMe_3)-(NHAr)Cl_4]$ is formed virtually quantitatively (119) is probably related to the reaction of $Re(CHCMe_3)(CH_2CMe_3)(NCMe_3)_2$ with three equivalents of 2,4-lutidine $\cdot$ HCl to form $[Re(CCMe_3)(CHCMe_3)(NH_2CMe_3)Cl_2]_2$ (116, 117). In both it is postulated that initial protonation of the imido ligand is followed by transfer of the neopentylidene α-proton to the nitrogen atom of either an imido ligand or an amido ligand. Subsequent protonations remove one of the nitrogen ligands as the ammonium salt. Cation exchange gives more soluble $[NEt_4]$-$[Re(CCMe_3)(NHAr)Cl_4]$ from which $Re(CCMe_3)(NAr)(OR)_2$ complexes are prepared virtually quantitatively.

$$Re(CHCMe_3)(NAr)_2Cl + 3\ HCl \xrightarrow{-78\,°C} [ArNH_3][Re(CCMe_3)(NHAr)Cl_4]$$

$$[Et_4N][Re(CCMe_3)(NHAr)Cl_4] + 3\ MOR \longrightarrow Re(CCMe_3)(NAr)(OR)_2$$
$$+ ROH + 2Et_4NCl + 3MCl$$

Scheme 45. M = Li, OR = $OCMe_3$, $OCMe_2(CF_3)$, $OCMe(CF_3)_2$, or DIPP (119); M = K, OR = $OC(CF_3)_2CF_2CF_2CF_3$(120).

$Re(CCMe_3)(NAr)[OCMe(CF_3)_2]_2$ reacts with two equivalents of 3-hexyne to yield an equivalent of $Me_3CC \equiv CEt$ and a rhenacyclobutadiene complex, $Re(C_3Et_3)(NAr)[OCMe(CF_3)_2]_2$ (119). The X-ray crystal structure of this metallacycle (Figure 8) shows that the ReC_3 ring is bent and unsymmetrically bound to the metal, reminiscent of $W[C(Ph)C(CMe_3)C(Ph)](\eta^5\text{-}C_5H_5)Cl_2$ (89). The dihedral angle between the C_α-Re-$C_{\alpha'}$ plane and C_α-C_β-$C_{\alpha'}$ plane of the metallacycle is 34° in $Re(C_3Et_3)(NAr)[OCMe(CF_3)_2]_2$; it is 56° in $W[C(Ph)C(CMe_3)C(Ph)](\eta^5\text{-}C_5H_5)Cl_2$. As in the tungsten case, the rhenacyclobutadiene ring is fluxional. The two α-ring carbon atoms interconvert, but the β-carbon atom remains distinct.

$Re(CCMe_3)(NAr)[OCMe(CF_3)_2]_2$ metathesizes dialkylalkynes in pentane at 25 °C at widely differing rates (5-undecyne > 4-nonyne > 3-heptyne). It is believed that metathesis occurs via rhenacyclobutadiene intermediates similar to $Re(C_3Et_3)(NAr)[OCMe(CF_3)_2]_2$ and that the rate is limited by loss of an alkyne from the rhenacyclobutadiene ring. Large alkynes are metathesized faster than small ones because rhenacycles containing small substituents are less labile. If the substituents are too large then rhenacycles are no longer stable, probably for steric reasons. For example, $Re(CCMe_3)(NAr)[OCMe(CF_3)_2]_2$ reacts with 2,5-dimethyl-3-hexyne to yield $Re(CCHMe_2)(NAr)[OCMe(CF_3)_2]_2$.

$Re(CSiMe_3)(CH_2SiMe_3)_3Cl$ has been isolated in $\sim 10\%$ yield from the reaction between $Re(thf)_2Cl_4$ and Me_3SiCH_2MgCl in THF (121). It was found to be a trigonal bipyramidal complex having an axial alkylidyne ligand with a Re ≡ C bond length of 172.6(11)pm and an axial chloride ligand. This bond length is virtually identical to that observed in $Re(CCMe_3)(CHCMe_3)I_2(py)_2$ (Table 1). The equatorial trimethylsilylmethyl ligands are bent away slightly from the axial trimethylsilylmethylidyne ligand, as one would expect on both steric and electronic grounds. These results illustrate once again that formally d^0 complexes that contain multiple metal-carbon bonds can be obtained in reactions that begin with starting materials in which the metal is not in its highest possible oxidation state.

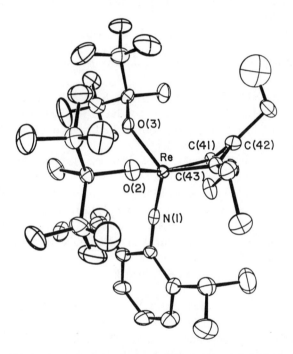

Figure 8. The structure of $Re(C_3Et_3)(N\text{-}2,6\text{-}C_6H_3Pr_2^i)[OCMe(CF_3)_2]_2$ (119); Re-C(41) = 188(1) pm; Re-C(43) = 209(1) pm; C(42)-C(43) = 133(2) pm; C(41)-C(42) = 146(1) pm; O(3)-Re-C(43) = 121.2°; O(3)-Re-N(1) = 119.1°; N(1)-Re-C(43) = 119.8°; O(2)-Re-C(41) = 141.8°; O(2)-Re-O(3) = 83.2°; O(3)-Re-C(41) = 105.5°.

5 References

(1) R. R. Schrock, Acc. Chem. Res. **19** (1986) 342–348.

(2) L. J. Guggenberger, R. R. Schrock, J. Am. Chem. Soc. **97** (1975) 2935.

(3) R. R. Schrock, Acc. Chem. Res. **12** (1979) 98–104.

(4) A. J. Schultz, J. M. Williams, R. R. Schrock, G. A. Rupprecht, J. D. Fellmann, J. Am. Chem. Soc. **101** (1979) 1593–1595.

(5) Transition Metal Carbene Complexes; Verlag Chemie: Weinheim, 1983.

(6) M. R. Churchill, F. J. Hollander, R. R. Schrock, J. Am. Chem. Soc. **100** (1978) 647–648.

(7) L. J. Guggenberger, R. R. Schrock, J. Am. Chem. Soc. **97** (1975) 6578–6579.

(8) F. Takusagawa, T. F. Koetzle, P. R. Sharp, R. R. Schrock, Acta. Cryst. B, in press.

(9) S. J. McLain, C. D. Wood, L. W. Messerle, R. R. Schrock, F. J. Hollander, W. J. Youngs, M. R. Churchill, J. Am. Chem. Soc. **100** (1978) 5962–5964.

(10) R. R. Schrock in Reactions of Coordinated Ligands; P. R. Braterman, Ed., Plenum: New York (1986).

(11) J. D. Fellmann, H. W. Turner, R. R. Schrock, J. Am. Chem. Soc. **102** (1980) 6608–6609.

(12) M. R. Churchill, W. J. Youngs, Inorg. Chem. **18** (1979) 171–176.

(13) R. R. Schrock, L. W. Messerle, C. D. Wood, L. J. Guggenberger, J. Am. Chem. Soc. **100** (1978) 3793–3800.

(14) J. D. Fellmann, G. A. Rupprecht, C. D. Wood, R. R. Schrock, J. Am. Chem. Soc. **100** (1978) 5964–5966.

(15) C. D. Wood, S. J. McLain, R. R. Schrock, J. Am. Chem. Soc. **101** (1979) 3210–3222.

(16) G. A. Rupprecht, L. W. Messerle, J. D. Fellmann, R. R. Schrock, J. Am. Chem. Soc. **102** (1980) 6236–6244.

(17) R. R. Schrock, J. Organomet. Chem. **300** (1986) 249–262.

(18) M. R. Churchill, H. J. Wasserman, H. W. Turner, R. R. Schrock, J. Am. Chem. Soc. **104** (1982) 1710–1716.

(19) F. Huq, W. Mowat, A. C. Skapski, G. Wilkinson, J. Chem. Soc., Chem. Commun. (1971) 1477–1478.

(20) R. A. Andersen, A. L. Galyer, G. Wilkinson, Angew. Chem. **88** (1976) 692–693; Angew. Chem., Int. Ed. Eng. **15** (1976) 609.

(21) P. E. Fanwick, A. E. Ogilvy, I. P. Rothwell, Organometallics **6** (1987) 73–80.

(22) A. W. Gal, H. J. Van der Heijden, J. Chem. Soc., Chem. Commun. (1983) 420–422.

(23) H. Van der Heijden, A.W. Gal, P. Pasman, A. G. Orpen, Organometallics **4** (1985) 1847–1853.

(24) D. N. Clark, R. R. Schrock, J. Am. Chem. Soc. **100** (1978) 6774–6776.

(25) R. R. Schrock, D. N. Clark, J. Sancho, J. H. Wengrovius, S. M. Rocklage, S. F. Pedersen, Organometallics **1** (1982) 1645–1651.

(26) L. G. McCullough, R. R. Schrock, J. C. Dewan, J. S. Murdzek, J. Am. Chem. Soc. **107** (1985) 5987–5998.

(27) R. A. Andersen, M. H. Chisholm, J. F. Gibson, W.W. Reichert, I. P. Rothwell, G. Wilkinson, Inorg. Chem. **20** (1981) 3934–3936.

(28) M. R. Churchill, W. J. Youngs, Inorg. Chem. **18** (1979) 2454–2458.

(29) J. S. Murdzek, Ph. D. Thesis, Massachusetts Institute of Technology (1988).

(30) C. J. Schaverien, R. R. Schrock, unpublished results.

(31) L. G. McCullough, Ph. D. Thesis, Massachusetts Institute of Technology (1984).

(32) M. R. Churchill, Y.-J. Li, J. Organomet. Chem. **282** (1985) 239–246.

(33) A. Mayr, G. A. McDermott, J. Am. Chem. Soc. **108** (1986) 548–549.

(34) S. F. Pedersen, R. R. Schrock, M. R. Churchill, H. J. Wasserman, J. Am. Chem. Soc. **104** (1982) 6808–6809.

(35) M. R. Churchill, H. J. Wasserman, J. Organomet. Chem. **270** (1984) 201–210.

(36) R. R. Schrock, S. F. Pedersen, M. R. Churchill, J.W. Ziller, Organometallics **3** (1984) 1574–1583.

(37) M. R. Churchill, J.W. Ziller, S. F. Pedersen, R. R. Schrock, J. Chem. Soc., Chem. Commun. (1984) 485–486.

(38) M. R. Churchill, H. J. Wasserman, Organometallics **2** (1983) 755–759.

(39) S. M. Rocklage, J. D. Fellmann, G. A. Rupprecht, L.W. Messerle, R. R. Schrock, J. Am. Chem. Soc. **103** (1981) 1440–1447.

(40) S. F. Pedersen, Ph. D. Thesis, Massachusetts Institute of Technology (1983).

(41) F. A. Cotton, W. Schwotzer, E. S. Shamshoum, Organometallics **3** (1984) 1770–1771.

(42) M. H. Chisholm, D. M. Hoffman, J. C. Huffman, Inorg. Chem. **22** (1983) 2903–2906.

(43) R. R. Schrock, M. L. Listemann, L. G. Sturgeoff, J. Am. Chem. Soc. **104** (1982) 4291–4293.

(44) M. L. Listemann, R. R. Schrock, Organometallics **4** (1985) 74–83.

(45) M. R. Churchill, J.W. Ziller, J. H. Freudenberger, R. R. Schrock, Organometallics **3** (1984) 1554–1562.

(46) J. H. Freudenberger, S. F. Pedersen, R. R. Schrock, Bull. Soc. Chim. France (1985) 349–352.

(47) J. H. Freudenberger, R. R. Schrock, M. R. Churchill, A. L. Rheingold, J.W. Ziller, Organometallics **3** (1984) 1563–1573.

(48) J. H. Wengrovius, J. Sancho, R. R. Schrock, J. Am. Chem. Soc. **103** (1981) 3932–3934.

(49) J. Sancho, R. R. Schrock, J. Mol. Catal. **15** (1982) 75–79.

(50) J. H. Freudenberger, R. R. Schrock, Organometallics **4** (1985) 1937–1944.

(51) S. L. Latesky, J. P. Selegue, J. Am. Chem. Soc. **109** (1987) 4731–4733.

(52) F. A. Cotton, W. Schwotzer, E. S. Shamshoum, Organometallics **2** (1983) 1167–1171.

(53) M. H. Chisholm, K. Folting, D. M. Hoffman, J. C. Huffman, J. Am. Chem. Soc. **106** (1984) 6794–6805.

(54) M. H. Chisholm, J. C. Huffman, N. S. Marchant, J. Am. Chem. Soc. **105** (1983) 6162–6163.

(55) Multiple Bonds Between Metal Atoms; F. A. Cotton, R. A. Walton, Eds. Wiley: New York (1982).

(56) I. A. Latham, R. R. Schrock, unpublished results.

(57) H. Strutz, R. R. Schrock, Organometallics **3** (1984) 1600–1601.

(58) B. E. Bursten, J. Am. Chem. Soc. **105** (1983) 121–122.

(59) R. H. Grubbs in Comprehensive Organometallic Chemistry; G. Wilkinson, F. G. A. Stone, E.W. Abel, Eds.; Pergamon: New York, 1982; Vol. 8, Chapter 54.

(60) T. J. Katz, J. M. McGinnis, J. Am. Chem. Soc. **97** (1975) 1592–1594.

(61) E. O. Fischer, U. Schubert, J. Organomet. Chem. **100** (1975) 59–81.

(62) F. Pennella, R. L. Banks, G. C. Bailey, J. Chem. Soc., Chem. Commun. (1968) 1548–1549.

(63) A. Mortreux, M. Blanchard, J. Chem. Soc., Chem. Commun. (1974) 786–787.

(64) A. Mortreux, J. C. Delgrange, M. Blanchard, B. Lubochinsky, J. Mol. Catal. **2** (1977) 73–82.

(65) S. Devarajan, D. R. M. Walton, G. J. Leigh, J. Organomet. Chem. **181** (1979) 99–104.

(66) M. Petit, A. Mortreux, F. Petit, J. Chem. Soc., Chem. Commun. (1982) 1385–1386.

(67) A. Bencheick, M. Petit, A. Mortreux, F. Petit, J. Mol. Catal. **15** (1982) 93–101.

(68) D. Villemin, P. Cadiot, Tetrahedron Lett. **23** (1982) 5139–5140.

(69) I. A. Latham, L. R. Sita, R. R. Schrock, Organometallics **5** (1986) 1508–1510.

(70) S. A. Krouse, Ph. D. Thesis, Massachusetts Institute of Technology (1988).

(71) J. H. Freudenberger, R. R. Schrock, Organometallics **5** (1986) 1411–1417.

(72) M. R. Churchill, J.W. Ziller, J. Organomet. Chem. **286** (1985) 27–36.

(73) L. G. McCullough, M. L. Listemann, R. R. Schrock, M. R. Churchill, J.W. Ziller, J. Am. Chem. Soc. **105** (1983) 6729–6730.

(74) M. R. Churchill, J.W. Ziller, J. Organomet. Chem. **281** (1985) 237–248.

(75) R. R. Schrock, J. S. Murdzek, J. H. Freudenberger, M. R. Churchill, J.W. Ziller, Organometallics **5** (1986) 25–33.

(76) H. Strutz, J. C. Dewan, R. R. Schrock, J. Am. Chem. Soc. **107** (1985) 5999–6005.

(77) S. A. Krouse, R. R. Schrock, R. E. Cohen, Macromolecules **20** (1987) 903–904.

(78) J. S. Murdzek, L. Blum, R. R. Schrock, Organometallics **7** (1988) 436–441.

(79) E. C. Walborsky, D. E. Wigley, E. Roland, J. C. Dewan, R. R. Schrock, Inorg. Chem. **26** (1987) 1615–1621.

(80) G. J. Leigh, M.T. Rahman, D. R. M. Walton, J. Chem. Soc., Chem. Commun. (1982) 541–542.

(81) M. L. Listemann, Ph. D. Thesis, Massachusetts Institute of Technology (1985).

(82) S. M. Rocklage, R. R. Schrock, M. R. Churchill, H. J. Wasserman, Organometallics **1** (1982) 1332–1338.

(83) J. S. Murdzek, R. R. Schrock, Organometallics **6** (1987) 1373–1374.

(84) C. J. Schaverien, J. C. Dewan, R. R. Schrock, J. Am. Chem. Soc. **108** (1986) 2771–2773.

(85) R. R. Schrock, R.T. DePue, J. Feldman, C. J. Schaverien, J. C. Dewan, A. H. Liu, J. Am. Chem. Soc. **110** (1988) 1423–1435.

(86) S. J. Holmes, R. R. Schrock, Organometallics **2** (1983) 1463–1464.

(87) M. R. Churchill, Y.-J. Li, L. Blum, R. R. Schrock, Organometallics **3** (1984) 109–113.

(88) M. R. Churchill, J.W. Ziller, L. McCullough, S. F. Pedersen, R. R. Schrock, Organometallics **2** (1983) 1046–1048.

(89) M. R. Churchill, J.W. Ziller, J. Organomet. Chem. **279** (1985) 403–412.

(90) M. R. Churchill, J. C. Fettinger, L. G. McCullough, R. R. Schrock, J. Am. Chem. Soc. **106** (1984) 3356–3357.

(91) M. R. Churchill, J. C. Fettinger, J. Organomet. Chem. **290** (1985) 375–386.

(92) A. H. Liu, R. C. Murray, J. C. Dewan, B. D. Santarsiero, R. R. Schrock, J. Am. Chem. Soc. **109** (1987) 4282–4291.

(93) A. H. Liu, Ph. D. Thesis, Massachusetts Institute of Technology (1988).

(94) J. H. Freudenberger, Ph. D. Thesis, Massachusetts Institute of Technology (1986).

(95) J. H. Freudenberger, R. R. Schrock, Organometallics **5** (1986) 398–400.

(96) K. Weiss, U. Schubert, R. R. Schrock, Organometallics **5** (1986) 397–398.

(97) J. Kress, A. Aguero, J. A. Osborn, J. Molec. Catal. **36** (1986) 1–35.

(98) J. Kress, M. Wesolek, J. A. Osborn, J. Chem. Soc., Chem. Commun. (1982) 514–516.

(99) J. Kress, J. A. Osborn, J. Am. Chem. Soc. **105** (1983) 6346–6347.

(100) I. Feinstein-Jaffe, S. F. Pedersen, R. R. Schrock, J. Am. Chem. Soc. **105** (1983) 7176–7177.

(101) I. Feinstein-Jaffe, D. Gibson, S. J. Lippard, R. R. Schrock, A. Spool, J. Am. Chem. Soc. **106** (1984) 6305–6310.

(102) I. Feinstein-Jaffe, J. C. Dewan, R. R. Schrock, Organometallics **4** (1985) 1189–1193.

(103) R. R. Schrock, S. Rocklage, J. H. Wengrovius, G. A. Rupprecht, J. D. Fellmann, J. Mol. Catal. **8** (1980) 73–83.

(104) J. H. Wengrovius, R. R. Schrock, M. R. Churchill, J. R. Missert, W. J. Youngs, J. Am. Chem. Soc. **102** (1980) 4515–4516.

(105) M. R. Churchill, A. L. Rheingold, W. J. Youngs, R. R. Schrock, J. H. Wengrovius, J. Organomet. Chem. **204** (1981) C17–C20.

(106) S. J. Holmes, Ph. D. Thesis, Massachusetts Institute of Technology (1983).

(107) K. Weiss, Angew. Chem. **98** (1986) 350–351; Angew. Chem., Int. Ed. Engl. **25** (1986) 359–360.

(108) S. J. Holmes, D. N. Clark, H.W. Turner, R. R. Schrock, J. Am. Chem. Soc. **104** (1982) 6322–6329.

(109) M. R. Churchill, A. L. Rheingold, H. J. Wasserman, Inorg. Chem. **20** (1981) 3392–3399.

(110) K.W. Chiu, R. A. Jones, G. Wilkinson, A. M. R. Galas, M. B. Hursthouse, K. M. A. Malik, J. Chem. Soc., Dalton Trans. (1981) 1204–1211.

(111) J. H. Wengrovius, R. R. Schrock, M. R. Churchill, H. J. Wasserman, J. Am. Chem. Soc. **104** (1982) 1739–1740.

(112) S. J. Holmes, R. R. Schrock, J. Am. Chem. Soc. **103** (1981) 4599–4600.

(113) S. J. Holmes, R. R. Schrock, M. R. Churchill, H. J. Wasserman, Organometallics **3** (1984) 476–484.

(114) A. J. Schultz, J. M. Williams, R. R. Schrock, S. J. Holmes, Acta Cryst. **C40** (1984) 590–592.

(115) M. R. Churchill, H. J. Wasserman, Inorg. Chem. **21** (1982) 3913–3916.

(116) D. S. Edwards, R. R. Schrock, J. Am. Chem. Soc. **104** (1982) 6806–6808.

(117) D. S. Edwards, L.V. Biondi, J.W. Ziller, M. R. Churchill, R. R. Schrock, Organometallics **2** (1983) 1505–1513.

(118) A. D. Horton, R. R. Schrock, J. H. Freudenberger, Organometallics **6** (1987) 893–894.

(119) R. R. Schrock, I. A. Weinstock, A. D. Horton, A. H. Liu, M. H. Schofield, J. Am. Chem. Soc. **110** (1988) 2686–2687.
(120) A. D. Horton, R. R. Schrock, unpublished results.
(121) P. D. Savage, G. Wilkinson, M. Motevalli, M. B. Hursthouse, Polyhedron **6** (1987) 1599–1601.

Catalytic Reactions
of Carbyne Complexes

Karin Weiss

1 Introduction

In 1976, only three years after the first synthesis of carbyne complexes (1) Fischer reported on the use of tungsten carbyne complexes as catalysts for the polymetathesis of cycloalkenes (2). With this work he started the career of carbyne complexes as catalysts. In the following years only few research groups studied the catalytic activities of carbyne complexes. This may be due to the fact, that the syntheses of Fischer type, as well as of Schrock type carbyne complexes are more complicated than the synthesis of Fischer type carbene complexes. Fischer type carbene complexes therefore have had the chance to develop as ideal model catalysts with low activity for the alkene metathesis reaction (3, 4, 5).

The most active alkene metathesis catalysts however are derived from reactions of transition metal compounds like tungsten or molybdenum halides or oxides with alkylating compounds like lithium-, aluminium- or tin-alkyls. From these reactions high valent tungsten or molybdenum carbene or carbyne complexes may result, as postulated by Dolgoplosk (6).

High-valent carbyne complexes of Schrock type should therefore be considered as important catalytic species for alkene metathesis reactions.

In addition catalytic alkyne metathesis and polymerization were studied in detail. Further catalytic reactions with well characterized carbyne complexes as catalysts or precatalysts have not yet been published.

2 Alkene Metathesis

2.1 Alkene Metathesis with Fischer-type Carbyne Complexes

In 1976 Fischer tested the catalytic alkene metathesis activity of the tungsten carbyne complexes $X(CO)_4W \equiv CPh$, $X = Cl$, Br, together with cocatalysts $(EtCl_2Al, TiCl_4, VCl_4, WCl_6$ and $SnCl_4)$ towards cycloalkenes (C_5, C_8, C_{12}) (Scheme 1).

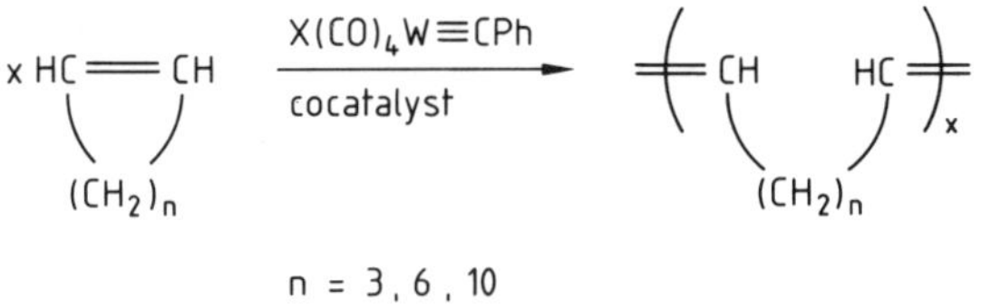

n = 3, 6, 10

Scheme 1.

In all cases the ring opening polymerization of the cycloalkenes, a metathesis reaction, gave polyalkenamers in high yield at $0°C$ within few minutes. The molar ratio carbyne complex : cycloalkene was from $1 : 10^4$ to $1 : 10^6$. The polyalkenamers had a predominantely trans configuration.

The corresponding chromium carbyne complexes gave no metathesis products. Similarly chromium carbene complexes do not catalyze metathesis reactions.

Together with Lewis acid cocatalysts the Fischer-type carbyne complexes proved to be very active metathesis catalysts for the ring opening polymerization of cycloalkenes. The reaction mechanism as well as the role of the cocatalysts in these metathesis reactions is still unknown. The cocatalysts probably act as halogenide acceptors forming highly reactive cationic coordinatively unsaturated carbyne complexes (Scheme 2).

$$X(CO)_4W \equiv CPh \; + \; MX_n \longrightarrow \; [(CO)_4W \equiv CPh]^{\oplus} \quad [MX_{n+1}]^{\ominus}$$

Scheme 2.

Similar activation of Schrock-type carbene complexes by metal halides was found by Kress and Osborn (7). They were able to isolate the cationic carbene complexes formed by the reaction with these cocatalysts.

Some of the cocatalysts used by Fischer might also function as oxidative, chlorinating agents to produce high valent carbyne complexes.

Some years later, in 1984, Katz tested the use of Fischer-type tungsten carbyne complexes as catalysts for ring opening polymerization of cycloalkenes without the cocatalysts employed by Fischer (8). But in addition he tested another cocatalyst which had already been successfully applied to metathesis reactions: oxygen (air) (Scheme 3).

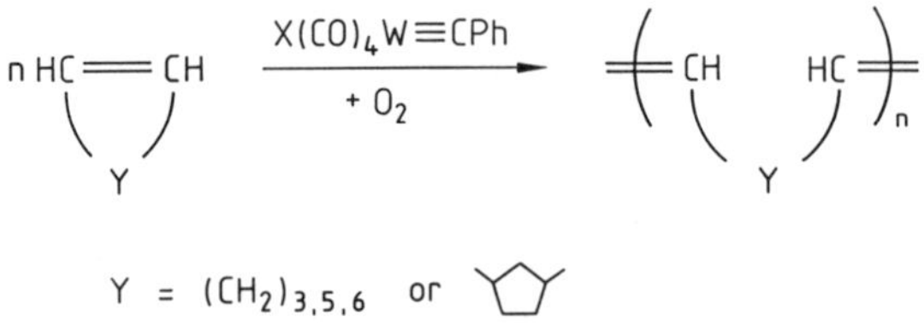

$$n\ HC = CH \underset{+\ O_2}{\overset{X(CO)_4W \equiv CPh}{\longrightarrow}} \left(= CH \qquad HC = \right)_n$$

Y (CH₂)₃,₅,₆

$$Y = (CH_2)_{3,5,6} \quad or \quad \bigcirc$$

Scheme 3.

Without any cocatalyst the very reactive norbornene was metathesized within few minutes in good yield. Cyclopentene, -heptene or -octene required reaction times of several hours at room temperature. It is known, however, that the Fischer-type carbyne complexes $X(CO)_4W \equiv CPh$, X = Cl, Br, I, used by Katz as catalysts decompose at room temperature. Therefore, after several hours at room temperature no starting carbyne complex would be present in the reaction mixture. With the more stable cyclopentadienyl-capped carbyne complex $Cp(CO)_2W \equiv CPh$ as catalyst Katz found no metathesis products, not even with norbornene.

The presence or absence of oxygen in the reaction mixture had a major effect on the yields of polyalkenamers. Addition of oxygen remarkably enhanced the yields of metathesis products. Changing the $C_{Carbyne}$ substituents from phenyl to pentachlorophenyl also increased the yields of metathesis products.

The product stereochemistry was analyzed by IR and ^{13}C NMR. The Fischer-type tungsten carbyne complexes alone, and together with oxygen, induced fairly high cis stereoselectivity. Only Fischer-type tungsten carbene complexes (9) alone, or together with phenylacetylene as cocatalyst (10) give polyalkenamers with higher cis-content.

For this reason, Katz suggested that Fischer-type carbyne complexes catalyze metathesis of alkanes because they act as sources of carbene complexes. He proposed that the trans-halogeno (tetracarbonyl) phenylcarbyne tungsten complexes rearrange to give 16 electron halogenocarbene complexes (Scheme 4).

$$Br(CO)_4W \equiv CPh \longrightarrow (CO)_4W = C \overset{Br}{\underset{Ph}{\diagdown}}$$

Scheme 4.

The resulting coordinatively unsaturated halogenocarbene complexes may coordinate an alkene to give tungsten cyclobutene derivatives and further metathesis.

The cocatalysts used by Fischer and Katz were able to activate Fischer-type tungsten carbyne complexes for the metathesis of cycloalkenes, but not for the metathesis of linear alkenes. An outstanding cocatalyst, the reduced Philips catalyst, activates Fischer-type carbyne complexes for the metathesis of linear alkenes (11).

The reduced Phillips catalyst, chromium(II) on a silica surface (12), is a coordinatively highly unsaturated compound which is used as polymerization catalyst for ethylene and for linear 1-alkenes (13). The reduced Phillips catalyst reacted in pentane suspension with Fischer-type carbyne complexes to form "heterogenized, activated carbyne complexes". The reaction is thought to proceed via a $[2 + 1]$ cycloaddition of the $W \equiv C_{carbyne}$ triple bond with the surface chromium(II) atoms to form bimetallacylopropene derivatives (Scheme 5).

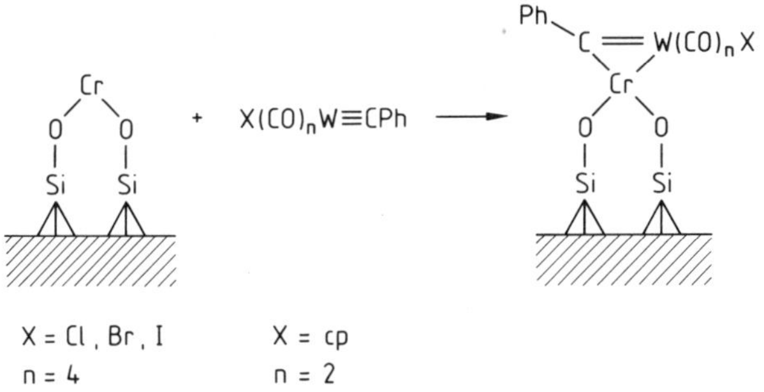

Scheme 5.

In these reactions the surface chromium atoms lost their polymerization activity for linear 1-alkenes, whilst the metathesis activity of the tungsten carbyne fragments were strongly activated. Even the bimetallic, heterogeneous catalyst formed from $Cp(CO)_2W \equiv CPh$ metathesizes 1-octene.

All bimetallic, heterogenous metathesis catalysts formed by reaction of Fischer-type carbyne complexes and reduced Phillips catalyst can be stored at $20\,°C$ for months without reduction of their metathetical activity, while the original Fischer-type carbyne complexes decompose as already mentioned.

The metathetical activities of the bimetallic catalysts were tested by reactions with 1-octene at $69\,°C$ in hexane (Scheme 6). The bimetallic catalysts with halogeno substituents X = Cl, Br, I, gave 80 % conversion of 1-octene within 2 hours, while the cyclopentadienyl substituted bimetallic catalyst required several hours (molar ratio carbyne: 1-octene = 1 : 100) (Scheme 6). The metathesis of 1-octene with the chlorosubstituted bimetallic catalyst was also tested in the molar ratio 1 : 1000 at $122\,°C$. Within 5 min metathesis equilibrium (60 % conversion of 1-octene) was achieved.

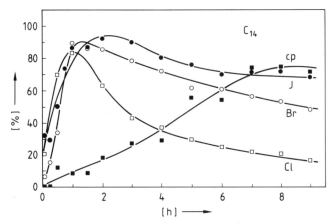

$$2 \quad CH_2{=}CHC_6H_{13} \quad \rightleftharpoons \quad CH_2{=}CH_2 \quad + \quad H_{13}C_6CH{=}CHC_6H_{13}$$
$$C_8 \qquad\qquad\qquad\qquad C_2 \qquad\qquad\qquad C_{14}$$

Scheme 6: Yields of Metathesis products C_{14} from Reactions of the bimetallic Catalysts $SiO_2/Cr/$ $W(CO)_n XCPh$ with 1-Octene. X = Cl, Br, I, n = 4; X = Cp, n = 2. Molar ratio W : 1-octene = 1 : 100.

The results of the three investigations of alkene metathesis with Fischer-type tungsten carbyne complexes as catalysts indicate that these carbyne complexes need cocatalysts to enhance their activity. Only the very reactive norbornene gives high yields without cocatalysts.

The role of the cocatalysts within the reaction pathway is not known. They may act as halogenide acceptors to form cationic, coordinatively unsaturated, carbyne complexes, or they may be able to chlorinate or oxidize the Fischer-type carbyne complexes to give high valent carbyne complexes. The elegant synthesis of Schrock-type carbyne complexes out of Fischer-type carbyne complexes by halogenation was found by Mayr in 1986 (14) and confirms the idea that Schrock-type carbyne complexes may be derived from Fischer-type carbyne complexes with metal halides as cocatalysts.

2.2 Alkene Metathesis with Schrock-type Carbyne Complexes

While the use of Fischer-type carbyne complexes as catalysts for alkene metathesis has been known since 1976, the emergence of Schrock-type carbyne complexes for alkene metathesis began in 1986.

The stoichiometric metathesis of C=N bonds of the heteroallenes (isocyanates (15) and carbodiimides (16)) with a Schrock carbyne complex $Cl_3(dme)W \equiv CCMe_3$ (dme = dimethoxyethane), led to the metathesis of a $W \equiv C_{carbyne}$triple bond with a C=N double bond (Scheme 7).

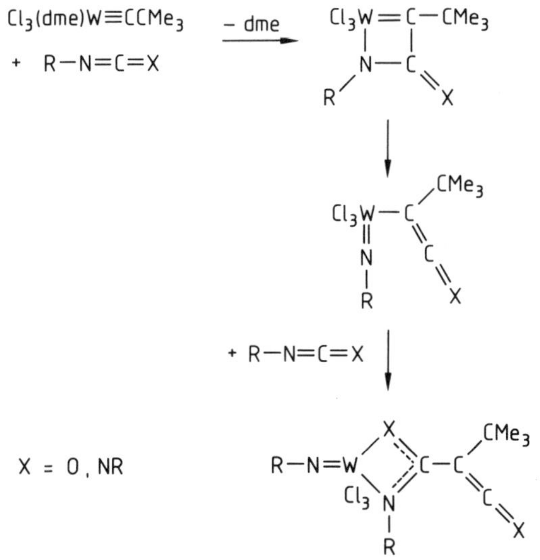

$$Cl_3(dme)W \equiv CCMe_3 \quad \xrightarrow{- \; dme}$$

$$+ \quad R-N=C=X$$

$$X = O, NR$$

Scheme 7.

The initially formed metathesis products reacted by insertion of a further molecule of heteroallene into the W-C bond, to form a chelating ligand.

Encouraged by this new type of metathesis reaction, experiments were started to test the metathesis of C=C double bonds of linear alkenes and cycloalkenes with this Schrock-type tungsten carbyne complex.

$Cl_3(dme)W \equiv CCMe_3$ proved to be a very reactive metathesis catalyst not only for cyacloalkenes, but also for 1-alkenes and internal alkenes (17). With cyclopentene 70 % polypentenamer was formed within few minutes. The molar mass was 70 000 compared with polystyrene and the trans-content of the polypentenamer was 75 %. Linear alkenes like 1-octene or 4-decene also gave metathesis with $Cl_3(dme)W \equiv CCMe_3$ in good yield within minutes. Addition of dme to the reaction mixture decreases the metathesis reaction rate drastically (18). This indicates that the dissoziation of dme might be the rate determining step of the metathesis.

The reaction is thought to be similar to that of the metathesis of heteroallenes (Scheme 8).

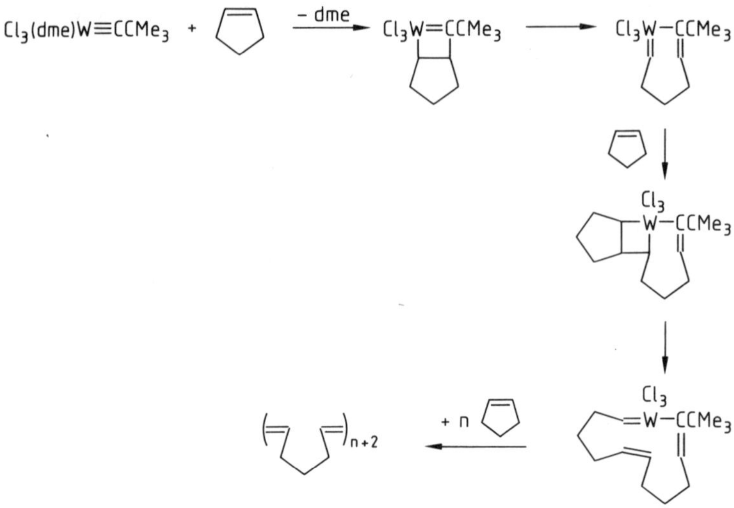

$$Cl_3(dme)W \equiv CCMe_3 \; + \; \bigcirc \quad \xrightarrow{-\,dme} \quad Cl_3W = CCMe_3 \quad \longrightarrow \quad Cl_3W - CCMe_3$$

Scheme 8.

The reaction of the alkene with the $W \equiv C_{carbyne}$ triple bond yields a tungsten cyclobutene derivative. Further reaction produces, by rearrangement, a carbene ligand and an alkyl ligand coordinated to tungsten. The following metathesis reactions occur on the $W = C_{carbene}$ double bond which is renewed with every metathesis cycle.

Alkyl substituents on the 1-alkenes on position C_4 to C_2 influence the metathesis increasingly (18). Addition of large amounts of 1-alkynes quenched alkene metathesis and yielded polyalkynes. The Schrock-type carbyne complex $(Bu^tO)_3W \equiv CCMe_3$, with bulky *t*-butoxy-ligands formed no metathesis products by reaction with linear alkenes or cycloalkenes.

In comparing the alkene metathesis activity of Fischer- and Schrock-type carbyne complex as catalysts, $Cl_3(dme)W \equiv CCMe_3$ proved to be the most active catalyst. This might be due to the fact, that this Schrock-type carbyne complex gives, by dissociation of dme, a coordinatively unsaturated complex, which favors the formation of a tungsten cylobutene by reaction with alkenes. Also the bulkiness of ligands may influence alkene metathesis. The coordinatively unsaturated $(Bu^tO)_3W \equiv CCMe_3$ complex with the bulky *t*-butoxy-ligands does not catalyze alkene metathesis, in contrast to the very active $Cl_3(dme)W \equiv CCMe_3$. However, it should also be pointed out, that the electrophilicity of the metal centre, or the metal carbon multiple bond is very important and electronic factors may also serve to deactivate the more electron-rich complex $(Bu^tO)_3W \equiv CCMe_3$.

The suggestion of Katz, that Fischer-type carbyne complexes act as sources of carbene complexes is also verified in a reaction pathway involving tungsten cyclobutene intermediates, as postulated for alkene metathesis with Schrock-type carbyne complexes. The cleavage of the tungsten cyclobutene gives a tungsten carbene complex on which further metathesis may occur.

The reason why the Schrock-type carbyne complex $Cl_3(dme)W \equiv CCMe_3$ is also able to metathize linear alkenes without cocatalysts and Fischer-type tungsten carbyne complexes do not, may again be due the fact, that $Cl_3(dme)W \equiv CCMe_3$ easily loses dme to form a tetrahedral 12 electron complex, while the dissociation of a CO ligand of the octahedral Fischer-type carbyne complexes requires more dissociation energy.

3 Alkyne Polymerization

Many alkene metathesis catalysts also induce catalytic reactions of alkynes (3). Generally 3 types of catalytic alkyne reactions are known (Scheme 9): a) trimerization, b) polymerization and c) metathesis of alkynes.

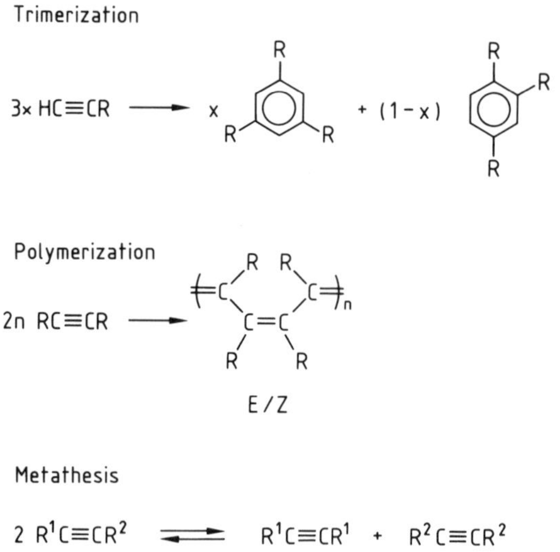

Trimerization

$$3\times HC\equiv CR \longrightarrow x \;\; \underset{R}{\overset{R}{\bigcirc}}_{R} \;\; +\; (1-x) \;\; \underset{R}{\overset{R}{\bigcirc}}{}^{R}$$

Polymerization

$$2n\; RC\equiv CR \longrightarrow \left[\!\!\left[C \underset{R}{\overset{R\;\;\;R}{\underset{C=C}{\diagup}}} C \right]\!\!\right]_n$$

E / Z

Metathesis

$$2\; R^1C\equiv CR^2 \;\; \rightleftharpoons \;\; R^1C\equiv CR^1 \;+\; R^2C\equiv CR^2$$

Scheme 9.

Whilst trimerization of alkynes with carbyne complexes was only observed as a side reaction, alkyne polymerization and alkyne metathesis are catalyzed by tungsten or molybdenum carbyne complexes.

3.1 Alkyne Polymerization with Fischer-type Carbyne Complexes

Katz tested Fischer-type carbene complexes (19) and Fischer-type carbyne complexes (8) as catalysts for alkyne polymerization (Scheme 10).

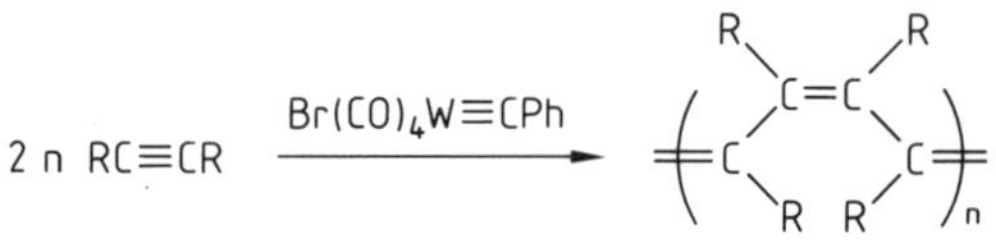

$$2\,n\; RC\equiv CR \xrightarrow{\;Br(CO)_4W\equiv CPh\;} \left[\!\!\left[C \underset{R\;\;\;R}{\overset{R\;\;\;\;\;R}{\underset{C}{\overset{C=C}{\diagup}}}} C \right]\!\!\right]_n$$

Scheme 10.

With the carbyne complex $Br(CO)_4W \equiv CPh$ mono- and disubstituted alkynes as well as alkynes with substituents bearing heteroatoms were polymerized (Table 1).

Table 1. Polymerization of alkynes with the Fischer-type carbyne complex $Br(CO)_4W \equiv CPh$ at 22 °C (according to reference 19)

Alkyne	Molar ratio alkyne: catalyst	Reaction Time [hours]	Yield Polymer	$M_w \cdot 10^{-3}$ (PS)
HC≡CH	536	92	34	–
HC≡CCH₃	380	8	5.2	14.4
HC≡CBuⁿ	200	48	6.4	120
HC≡CBuᵗ	190	48	39	180
HC≡CPh	151	3	10	151
	151	14	63	–
CH₃C≡CCH₃	350	19	12	–
CH₃C≡CC₂H₅	209	20	20	296
HC≡C(CH₂)₃Cl	218	2.5	53	201
HC≡C(CH₂)₃CO₂CH₃	147	38	16	14
HC≡C(CH₂)₃CN	80	48	16	10

If the alkynes bear substituents with heteroatoms (like $C \equiv N$, COOMe, OMe) the yields of polyacetylenes formed were only satisfactory if the substituents were situated on C_n ($n \geq 5$) as for example in $HC \equiv C(CH_2)_3\text{-}C \equiv N$.

The stereochemistry of the polymers were analyzed by IR, ^{13}C NMR and ^1H NMR. The polymerization of the unsubstituted $HC \equiv CH$ with $Br(CO)_4W \equiv CPh$ gives insoluble trans-polyacetylene. In contrast polyphenylacetylene and poly-*t*-butylacetylene were shown to have a predominantely cis-(E) configuration.

Katz also tested two further carbyne complexes. The methylcarbyne complex $Br(CO)_4W \equiv CMe$ reacts with propyne at –15 °C to give 11 % polypropyne after 11 days. The cyclopentadienyl capped $Cp(CO)_2W \equiv CPh$ formed only 1.6 % polymer on reaction with phenylacetylene.

For the reaction mechanism Katz suggested that alkyne polymerization and alkene metathesis with Fischer-type transhalogeno carbyne complexes start by a rearrangement of the carbyne complexes to give halogenocarbene complexes.

These coordinatively unsaturated carbene complexes may react with alkynes to give tungsten cyclobutene derivatives. A ring opening reaction forms a new carbene complex. The carbene ligand contains the growing polyalkyne (Scheme 11).

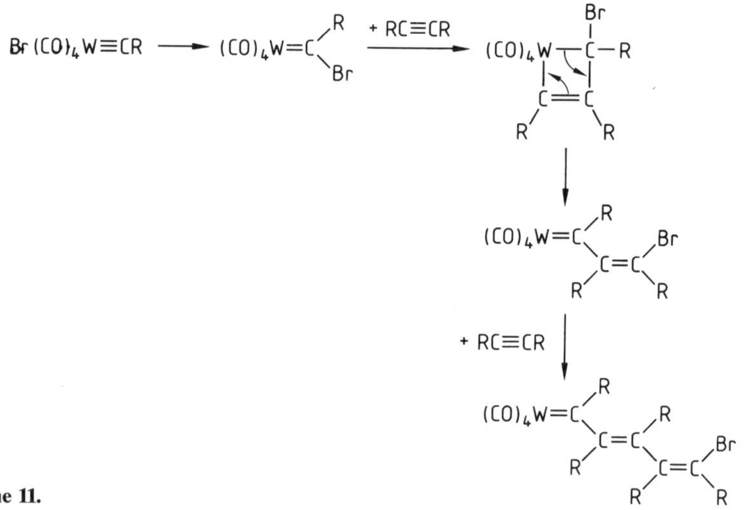

Scheme 11.

The observation that the cyclopentadienyl capped carbyne complex $Cp(CO)_2W \equiv CPh$ is not an effective catalyst for alkyne polymerization (or alkene metathesis) agrees with this hypothesis. Also the fact that the stereochemistry of the polyalkynes produced with the Fischer-type carbyne complex $Br(CO)_4W \equiv CPh$ and the carbene complex $(CO)_5W = C(OMe)Ph$ are very similar, accords with the idea that in both cases carbene complexes are the catalytically active species.

3.2 Alkyne Polymerization with Schrock-type Carbyne Complexes

Schrock tested many stoichiometric and catalytic reactions with alkynes of tungsten(VI) carbyne complexes (20, 21). But the catalytic reactions were focussed on alkyne metathesis (Chapter 4). Only when he tested molybdenum(VI) carbyne complexes as catalysts for alkyne metathesis, did he find that $(Pr^iO)_3$-$MoCCMe_3$ and $(Me_3CCH_3O)MoCCMe_3$ polymerize 3-heptyne (22).

That tungsten(VI) carbyne complexes act as polymerization catalysts for alkynes, was clear from the tests on the influence of alkynes on alkene metathesis with $Cl_3(dme)W \equiv CCMe_3$ (18) (Chapter 2.2). Three differently substituted tungsten(VI) carbyne complexes were tested to study the alkyne polymerization activity: $Cl_3(dme)W \equiv CCMe_3$, $(Bu^tO)_3W \equiv CCMe_3$ and $(Me_3CCH_2)_3$ $W \equiv CCMe_3$ (18, 23). All three Schrock-type carbyne complexes proved to be active polymerization catalysts for alkynes (Table 2).

Table 2. Polymerization of alkynes with differently substituted Schrock-type tungsten carbyne complexes (in CH_2Cl_2 at 20°C, according to references 18 and 23)

Catalyst	Alkyne	Molar ratio alkyne: catalyst	Yield of Polymer [%]	$M_w \cdot 10^{-3}$ (PS)
$Cl_3(dme)W \equiv CCMe_3$	$HC \equiv CPh$	63	69	32
	$HC \equiv CCMe_3$	99	82	300
	l-hexyne	118	12	3.8
	l-octyne	54	28	2.8
	4-octyne	44	32	1.2
$(Bu^tO)_3W \equiv CCMe_3$	$HC \equiv CPh$	90	93	80
	$HC \equiv CPh$	190	94	220
	$HC \equiv CCMe_3$	99	86	200
	l-hexyne	120	91	16
	l-octyne	120	90	110
	4-octyne	30	28	2
$(Bu^tCH_2)_3W \equiv CCMe_3$	$HC \equiv CPh$	110	77	16
	$HC \equiv CCMe_3$	110	64	133
	l-octyne	100	91	61
	4-octyne	100	6	0.4

The resulting polyphenylacetylene and poly-*t*-butylacetylene have predominantly cis-configuration, as in the polymerization with Fischer-type carbene and carbyne complexes as catalysts. The cis-content increased with the catalyst order $Cl_3(dme)W \equiv CCMe_3$ < $(Bu^tO)_3W \equiv CCMe_3$ < $(Bu^tCH_2)_3$ $W \equiv CCMe_3$. *t*-Butylacetylene and phenylacetylene were polymerized within a few minutes with high yields. Addition of dme to the reaction mixtures with $Cl_3(dme)W \equiv CCMe_3$ as catalyst caused a delay but no inhibition of the polymerization. This indicates that the dissociation of the dme ligand could be the rate determining step of the polymerization. n-Alkyl-substituted alkynes like 1-hexyne, 1-octyne or 4-octyne give lower yields and lower molar masses of the polymeric products with increasing alkyl-substituents. Together with the polymeric methanol insoluble reaction products, some oligomeric methanol soluble products formed. All methanol soluble fractions contained small amounts of trimerization products of the alkynes.

Schrock had already studied stoichiometric reactions of internal alkynes with $Cl_3(dme)W \equiv CCMe_3$ (24, 25). He was able to isolate tungsten cyclobutadiene derivatives which react with a further mole of alkyne to give cyclopentadienyl-complexes via tungstena benzene derivatives which were not isolated. The formation of tungstena benzene derivatives may also represent the first steps of the alkyne polymerization. If there is a high concentration of alkynes present, the formation of cyclopentadienyl complexes is not favored and further polymerization occurs (Scheme 12).

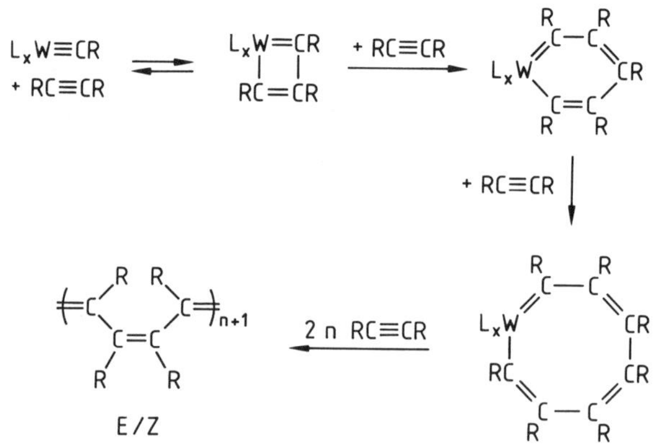

Scheme 12.

The three Schrock-type carbyne complexes polymerized not only 1-alkynes but also 4-octyne, (an internal alkyne), however with lower yields. Therefore the polymerization mechanism presented in Scheme 12 seems to be more likely than a mechanism which starts with a linear carbene ligand. Carbene complexes can be formed as by-products by reactions of carbyne complexes with 1-alkynes. A vinyl-substituted carbene complex was isolated by reaction of $CpCl_2W \equiv CCMe_3$ with t-butylacetylene together with a dehydro tungstena cyclobutadiene derivative (26) (Scheme 13). Internal alkynes without acidic hydrogens, unlike 1-alkynes, cannot form such vinylsubstituted carbene complexes. Therefore an alkyne polymerization mechanism proceeding via linear carbene complexes is unlikely.

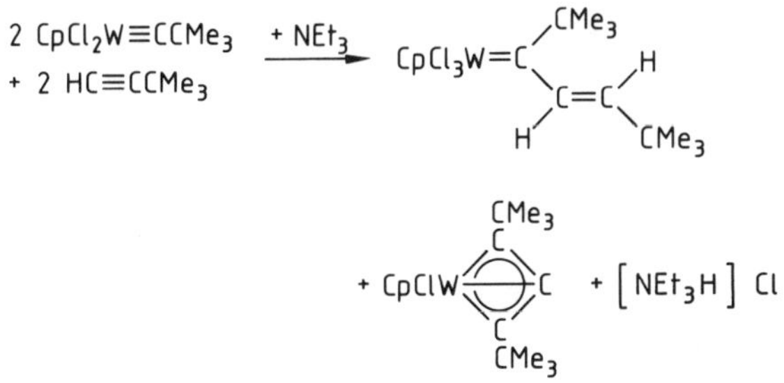

$$2\ CpCl_2W\equiv CCMe_3\quad +\ NEt_3$$
$$+\ 2\ HC\equiv CCMe_3$$

$$\longrightarrow\quad CpCl_3W=C\Big\langle {}^{CMe_3}_{C=C\langle{}^H_{CMe_3}} \Big.$$

$$+\ CpClW\ \langle\ \rangle C\ \ +\ \big[NEt_3H\big]\ Cl$$

Scheme 13.

The fact that the carbyne complex $(Bu^tO)_3W\equiv CCMe_3$ is an active alkyne polymerization catalyst for 1-alkynes and internal alkynes was very surprising, because Schrock had found that this catalyst is an active metathesis catalyst for internal alkynes (27) (see Chapter 4).

The results of alkyne polymerizations with Fischer-type and Schrock-type tungsten carbyne complexes show that both are active catalysts, especially for 1-alkynes. Internal alkynes react with both types of catalysts to give polymeric or oligomeric reaction products. The poly-*t*-butylacetylene and polyphenylacetylene formed were predominately in the cis-(E) configuration.

For the reaction pathway two different mechanisms had been suggested. Polymerization of alkynes by Fischer-type carbyne complexes is thought to be started by carbene complexes, which are formed by rearrangement of the carbyne complexes. Schrock-type carbyne complexes may give in the first reaction metallacyclobutadiene derivatives, which also represent carbene complexes and these are the active species for further reactions with alkynes. The reaction mechanism proposed for alkyne polymerization with Schrock-type carbyne complexes is theoretically also possible for Fischer-type carbyne complexes but the low valent tungsten cyclobutadiene derivatives to support this idea are hitherto unknown. However many tungsten(VI)cyclobutadiene derivatives were isolated (Chapter 4).

4 Alkyne Metathesis with Schrock-type Carbyne Complexes

1968 the first catalytic alkyne metathesis was found by Panella, Banks and Bailey (28). They used the heterogeneous alkene metathesis catalyst WO_3/SiO_2 for the metathesis reaction of 2-pentyne at 350 °C and isolated 2-butyne and 3-hexyne (Scheme 14).

$$2 \ CH_3C{\equiv}CCH_2CH_3 \ \xrightleftharpoons[350\,°C]{WO_3/SiO_2} \ CH_3C{\equiv}CCH_3 \ + \ CH_3CH_2C{\equiv}CCH_2CH_3$$

Scheme 14.

In addition to alkyne metathesis products other products with higher molar mass were formed. Boelhouwer used this catalyst to study the metathesis reactions of numerous 1-alkynes and found that cyclotrimerizations predominated to give substituted benzene derivatives along with some metathesis of the 1-alkynes (29). MoO_3/SiO_2, especially in combination with cobalt oxides, was also found to be an active catalyst for 1-alkyne metathesis st 300–500 °C (30, 31).

Many homogeneous alkyne metathesis catalysts have been found in the meantime (3, 4). With a homogeneous catalyst, formed by the reaction of $Mo(CO)_6$ with phenolic compounds Montreux, in 1978, was able to show, by use of ^{13}C labelled alkyne, that alkyne metathesis proceeds via alkylidyne (= carbyne) group exchange (32) (Scheme 15)

$$2 \ Ph-C^{*}{\equiv}C-Bu \ \xrightleftharpoons{} \ Ph-C^{*}{\equiv}C^{*}-Ph \ + \ Bu-C{\equiv}C-Bu$$

Scheme 15.

Some years earlier Katz had already postulated that carbyne complexes are the catalytic species in alkyne metathesis reactions (33) (Scheme 16).

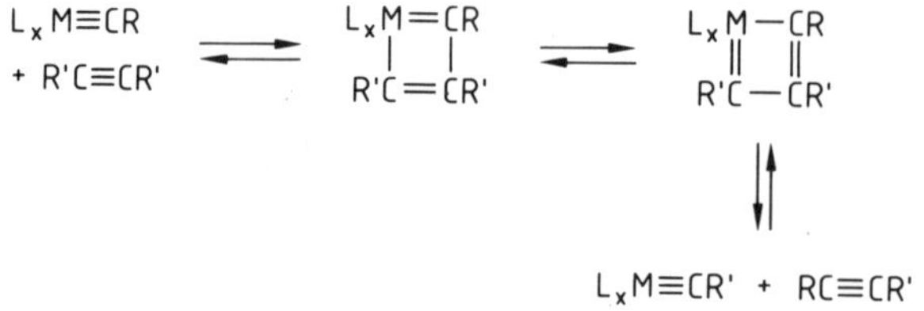

$$
\begin{array}{ccc}
L_xM{\equiv}CR & L_xM{=}CR & L_xM{-}CR \\
+ \ R'C{\equiv}CR' & R'C{=}CR' & R'C{-}CR'
\end{array}
$$

$$L_xM{\equiv}CR' \ + \ RC{\equiv}CR'$$

Scheme 16.

With Fischer-type carbyne complexes no alkyne metathesis had hitherto been observed. However, in 1981 Schrock was able to demonstrate that tungsten(VI) (27, 35) and molybdenum(VI) (22) carbyne complexes give not only stoichiometric, but also catalytic metathesis of alkynes.

The ability to catalyze alkyne metathesis via an exchange of carbyne ligands represents the most specific of all catalytic reactions of carbyne complexes. $(Bu^tO)_3W \equiv CCMe_3$ was the first carbyne complex which was successfully used as an alkyne metathesis catalyst (27). With this catalyst, not only metathesis of differently alkyl and aryl disubstituted alkynes was achieved, but also metathesis of alkynes with substituents bearing heteroatoms. Kinetic studies revealed that the metathesis reactions are first order in tungsten and in alkyne.

Katz had postulated metallacyclobutadiene derivatives as intermediates in alkyne metathesis reactions (Scheme 16). In the reaction mixtures of $(Bu^tO)_3W \equiv CCMe_3$ with alkynes, no tungstena cyclobutadienes were found, but the chloro-substituted carbyne complex $Cl_3(dme)W \equiv CCMe_3$ yielded tungstena cyclobutadiene derivatives in stoichiometric reactions with internal alkynes with a pseudo-trigonal bipyramidal structure and a planar ring in the equatorial plane (24, 36) (Scheme 17).

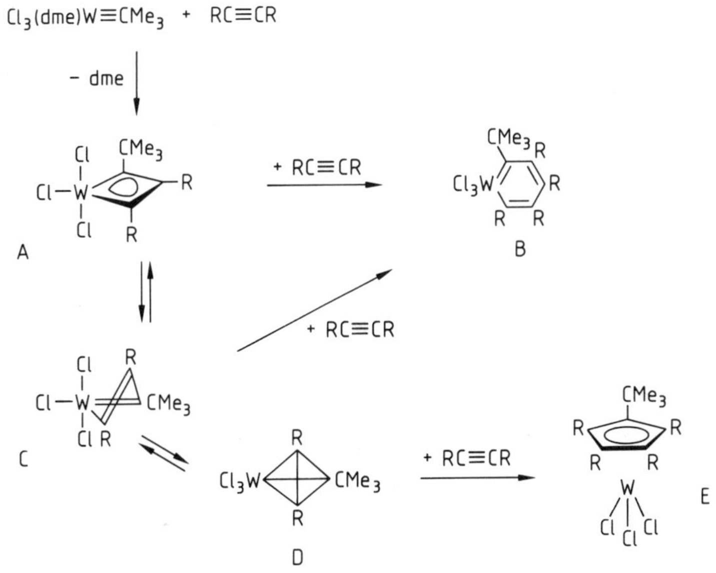

Scheme 17.

The planar tungstena cyclobutadiene derivative A reacts with a further molecule of alkyne to give the tungstena benzene derivative B. If the auxiliary ligands are neither bulky enough nor electron withdrawing, the planar tungstena cyclobutadiene A reacts via a nonplanar fluxional cyclobutadiene C to give a η^3-cyclopropenyl tungstena complex D (37).

Such metalla-cyclopropenyl (or metalla-tetrahedrane) complexes can be stabilized by additional donor ligands, like pyridine or tetramethylethylendiamine. With additional alkyne, D forms a stable η^5-cyclopentadienyl tungsten complex E. If the chloro ligands in the planar cyclobutadiene complex A are replaced by bulky alkoxy ligands, alkyne metathesis products result via a retro [2 + 2] cycloaddition (36).

Tungsten carbyne complexes yield no metathesis products by reaction with 1-alkynes. With stoichiometric amounts of 1-alkynes dehydro tungstena cyclobutadiene complexes are formed (26) (Scheme 13). An ecxess of 1-alkynes gives polymerization of 1-alkynes (see Chapter 3.2).

In contrast to the tungsten carbyne complex $(Bu^tO)_3W \equiv CCMe_3$, which is a very active metathesis catalyst for internal alkynes, the analogous molybdenum complex $(Bu^tO)_3Mo \equiv CCMe_3$ gives only unidentified polymeric products with internal alkynes. $(Bu^tCH_2O)_3Mo \equiv CCMe_3$ and $(Pr^iO)_3Mo \equiv CCMe_3$ both polymerize alkynes. Molybdenum carbyne complexes with more electron withdrawing alkoxy substituents R (R = $OC(CF_3)_3$, $OCH(CF_3)_2$, $OCMe_2(CF_3)$, $OCMe(CF_3)_2$ or $0-2,6-C_6H_3(Pr^i)_2$), or with alkoxy ligands partly replaced by chloro ligands, give complexes which are active alkyne metathesis catalysts (38, 22, 39) (Table 3) for internal alkynes. With 1-alkynes these catalysts give dehydromolybdena-cyclobutadiene derivatives. With 1-alkynes $(Bu^tO)_3Mo \equiv CCMe_3$ reacts in a stoichiometric metathesis reaction (38) (Scheme 18).

$$(Me_3CO)_3 \, Mo \equiv CCMe_3 \quad \longleftrightarrow \quad (Me_3CO)_3 \, Mo \equiv CR$$

$$+ \; RC \equiv CH \qquad\qquad\qquad + \; Me_3C \equiv CH$$

$$R = Ph \, , \; CHMe_2$$

Scheme 18.

Tungsten carbyne complexes, with alkoxy ligands substituted with more electron withdrawing substituents similarly to molybdenum carbyne complexes, react with internal alkynes to yield planar tungstena cyclobutadiene derivatives, which, with excess of alkynes, give metathesis products (25).

The studies by Schrock on alkyne metathesis with carbyne complexes has yielded many results that elucidate the reaction pathway of alkyne metathesis.

Table 3. Metathesis of alkynes with Schrock-type tungsten and molybdenum carbyne complexes (according to references 25, 27, 35–39)

Catalyst	Alkyne	Molar ratio alkyne: catalyst	Time to metathesis equilibrium
$(Bu^tO)_3W{\equiv}CCMe_3$	$PhC{\equiv}CEt$	19	4 h
	$PhC{\equiv}CTol$	30	1 h
	$CyC{\equiv}CEt$	27	18 min
	$EtC{\equiv}CPr$	2.300	10 min
$[(CF_3)_3CO]_3dmeMo{\equiv}CCMe_3$	$EtC{\equiv}CPr$	20	<1 min
$[Me_2(CF_3)CO]Mo{\equiv}CCMe_3$	$EtC{\equiv}CPr$	20	30 min
$[(CF_3)_2CHO]_3dmeMo{\equiv}CCMe_3$	$EtC{\equiv}CPr$	20	<5 min
$(RO)_3W{\equiv}CCMe_3$	$EtC{\equiv}CPr$	20	4.5 h
$R=0-2,6C_6H_3(Pr^i)_2$	$PrC{\equiv}CBu$	20	25 min

He was able to isolate numerous molybdena- and tungstena-cyclobutadiene derivatives, and some were active alkyne metathesis catalysts. He found that bulky alkoxy ligands are especially desirable for metathesis activity, for both molybdenum and tungsten carbyne complexes. More electron withdrawing substituents also favour alkyne metathesis. As regards the occasionally different reactivity of comparable molybdenum and tungsten carbyne complexes both Schrock and Chisholm (40) suggested that molybdenum carbyne complexes are less electrophilic than those of tungsten. Another condition for tungsten and molybdenum carbyne complexes to be active alkyne metathesis catalysts is that they form with alkynes metalla-cyclobutadiene derivatives with planar WC_3 rings. Metalla-cyclobutadienes such as $[0-2,5-C_6H_3(Pr^i)_2]_3WC_3Et_3$ metathize alkynes like 3-hexyne d_{10}. Kinetic studies showed that the reaction is first order in the metalla-cycle and zero order in the 3-hexyne d_{10}. The results support a mechanism in which the rate determining step is the cleavage of the metalla-cyclobutadiene to give an alkyne and a new carbyne complex (41).

This cleavage of the metalla-cyclobutadienes is favored by both electron withdrawing and bulky substituents on the molybdenum and tungsten atoms, but not by the addition of a further alkyne to a $W = C_{carbene}$ double bond (Scheme 19).

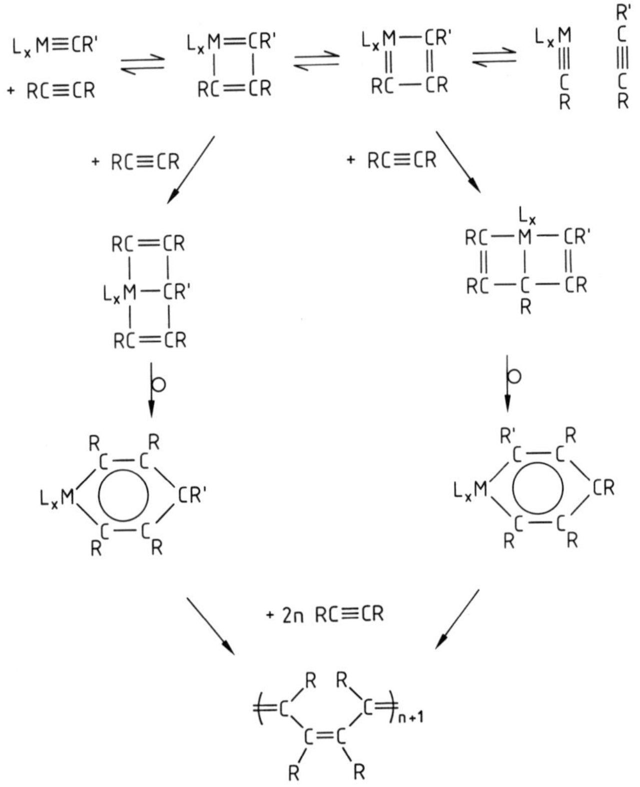

Scheme 19.

Metalla-cyclobutadiene derivatives are cyclic carbene complexes. Hence they are expected to react with alkynes like Fischer-type (42) or Schrock-type (43) carbene complexes to give polyalkynes.

Whether carbyne complexes give metathesis of alkynes or not, depends on the reactivity of the metalla-cyclobutadiene derivatives. If steric and electronic factors favour the reaction of the metalla-cyclobutadiene with the alkyne, polymerization occurs. Such factors include smaller ligands which have little electron withdrawing effect on the electron density of the $M \equiv C_{carbyne}$ triple bond.

5 Conclusions

The catalytic reactions of carbyne complexes which are known, are all concentrated on one basic reaction type: a $[2 + 2]$ cyclo-addition reaction of alkenes or alkynes to the $M \equiv C_{carbyne}$ triple bond. These cyclo-additions are very similar to the numerous cyclo-additions of alkenes or alkynes to the $M = C_{carbene}$ double bond of carbene complexes (3–5) (Scheme 20).

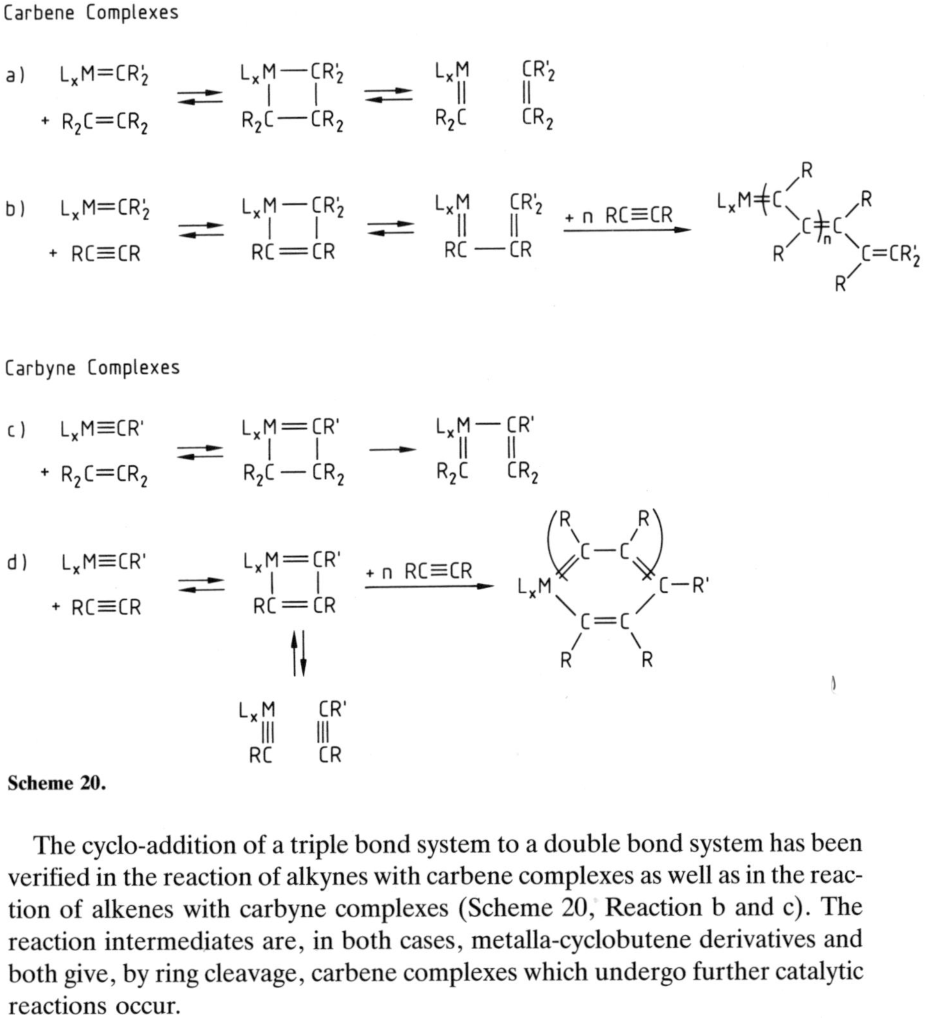

Carbene Complexes

a) $L_xM = CR'_2$ $\rightleftharpoons$ $L_xM - CR'_2$ $\rightleftharpoons$ $L_xM \quad CR'_2$
     $+ \ R_2C = CR_2$ $R_2C - CR_2$ $R_2C \quad CR_2$

b) $L_xM = CR'_2$ $\rightleftharpoons$ $L_xM - CR'_2$ $\rightleftharpoons$ $L_xM \quad CR'_2$ $+ \ n \ RC \equiv CR$
     $+ \ RC \equiv CR$ $RC = CR$ $RC - CR$

Carbyne Complexes

c) $L_xM \equiv CR'$ $\rightleftharpoons$ $L_xM = CR'$ $\longrightarrow$ $L_xM - CR'$
     $+ \ R_2C = CR_2$ $R_2C - CR_2$ $R_2C \quad CR_2$

d) $L_xM \equiv CR'$ $\rightleftharpoons$ $L_xM = CR'$ $+ \ n \ RC \equiv CR$ $\longrightarrow$
     $+ \ RC \equiv CR$ $RC = CR$

             $\updownarrow$

             $L_xM \quad CR'$
             $RC \quad CR$

Scheme 20.

The cyclo-addition of a triple bond system to a double bond system has been verified in the reaction of alkynes with carbene complexes as well as in the reaction of alkenes with carbyne complexes (Scheme 20, Reaction b and c). The reaction intermediates are, in both cases, metalla-cyclobutene derivatives and both give, by ring cleavage, carbene complexes which undergo further catalytic reactions occur.

The metathesis reactions of alkenes with carbene complexes and of alkynes with carbyne complexes (Scheme 20, Reaction a and d) are also very similar reactions according to this point of view. Both form cyclic intermediates via a [2 + 2] cycloaddition reaction, metalla-cyclobutane or metalla-cyclobutadiene derivatives. A retro [2 + 2] cycloaddition in both cases yields the metathesis products. The metalla-cyclobutadienes formed by reaction of alkynes with carbyne complexes are carbene complexes and may, in addition, therefore give rise to alkyne polymerization.

With one exception, all catalytic reactions which are outlined in Scheme 20 are known for both Fischer-type carbyne complexes (with low-valent transition metals) and Schrock-type carbyne complexes (with high valent transition metals). The unknown reaction is the metathesis of alkynes with Fischer-type carbyne complexes.

The reaction intermediates of all cyclo-additions do not appear to depend on whether an electrophilic or nucleophilic transition metal is involved.

However, the catalytic activity depends remarkably on the fact whether the transition metal of the carbyne complex is nucleophilic or electrophilic. The high valent Schrock-type carbyne complexes with electrophilic transition metals proved to be catalytically more active.

Stoichiometric reactions of heteroalkenes and heteroallenes with Schrock-type carbyne complexes (15, 23) revealed that metathesis via [2 + 2] cycloaddition reactions are not limited to C,C double or triple bond systems of alkenes and alkynes. Recently, catalytic metathesis of imines and carbodiimides with Schrock-type carbyne complexes was developed (23). The catalytic imine metathesis reaction might be of interest for biochemical reactions.

References

(1) E. O. Fischer, G. Kreis, C. G. Kreiter, J. Müller, G. Huttner and H. Lorenz, Angew. Chem. **85** (1973) 618, Angew. Chem. Int. Ed. Engl. **12** (1973) 564.

(2) a) E. O. Fischer and W. R. Wagner, J. Organomct. Chem. **116** (1976) C21
b) W. R. Wagner, Thesis Techn. Universität München 1978

(3) K. J. Ivin, Olefin Metathesis, Academic Press, London 1983

(4) V. Dragutan, A. T. Balaban und M. Dimonie, Olefin Metathesis and Ring-opening Polymerisation of Cycloolefins, John Wiley Ed. New York 1985

(5) K. Weiss in K. H. Dötz, Transition Metal Carbene Complexes, Verlag Chemie 1983

(6) B. A. Dolgoplosk, J. Mol. Cat. **15** (1982) 193

(7) J. Kress, A. Aguero and J. A. Osborn, J. Mol. Cat. **36** (1986) 1

(8) T. J. Katz, T. H. Ho, N. Y. Shih, Y. C. Ying and V. I. Stuart, J. Am. Chem. Soc. **106** (1984) 2659

(9) T. J. Katz, S. J. Lee and N. Acton, Tetrahedron Lett. (1976) 4247

(10) T. J. Katz, S. J. Lee, N. Nair and E. B. Savage, J. Am. Chem. Soc. **102** (1980) 7940

(11) K. Weiss and M. Denzner, J. Organomet. Chem. **355** (1988) in press

(12) a) H. L. Krauss, B. Rebenstorf and U. Westphal, Z. Anorg. Allgem. Chem. **414** (1975) 97
b) M. P. McDaniel, Advances in Catal. **33** (1985) 47

(13) K. Weiss and H. L. Krauss, J. Catal. **88** (1984) 424

(14) A. Mayr and G. A. McDermott, J. Am. Chem. Soc. **108** (1986) 548

(15) K. Weiss, U. Schubert and R. R. Schrock, Organometallics, **5** (1986) 397

(16) K. Weiss, unpuplished results

(17) K. Weiss, Angew. Chem. 98 (1986) 350, Angew. Chem. Int. Ed Engl. **25** (1986) 359

(18) K. Weiss, R. Goller and G. Lößel, J. Mol. Cat. **46** (1988) 267

(19) T. J. Katz and S. J. Lee, J. Am. Chem. Soc. **102** (1980) 422

(20) R. R. Schrock, J. H. Freudenberger, M. L. Listeman and L. G. McCullough, J. Mol. Cat. **28** (1985) 1

(21) R. R. Schrock, J. Organomet. Chem. **300** (1986) 249

(22) L. G. McCullough and R. R. Schrock, J. Am. Chem. Soc. **106** (1984) 4067

(23) R. Goller, Thesis, Universität Bayreuth, 1988

(24) S. F. Pedersen, R. R. Schrock, M. R. Churchill and H. J. Wassermann, J. Am. Chem. Soc. **104** (1982) 6808

(25) J. H. Freudenberger, R. R. Schrock, M. R. Churchill, A. L. Rheingold and J. W. Ziller, Organometallics **3** (1984) 1563

(26) L. G. McCullough, M. L. Listeman, R. R. Schrock, M. R. Churchill and J.W. Ziller, J. Am. Chem. Soc. **105** (1983) 6729

(27) J. H. Wengrovius, J. Sancho and R. R. Schrock, J. Am. Chem. Soc. **102** (1981) 3932

(28) F. Panella, R. L. Banks and C. G. Bailey, J. Chem. Soc. Chem. Commun. (1968) 1548
USP 3655804 (1968) Phillips Petrol Co.

(29) J. A. Moulijn, H. J. Reitsma and C. Boelhouwer, J. Catal. **25** (1972) 434

(30) A.V. Mushegyan, Arm. Khim. Zh. 28 (1975) 672 Chem. Abstr. **84** (1976) 16675

(31) A.V. Mushegyan, Arm. Khim. Zh. 28 (1975) 674 Chem. Abstr. **84** (1976) 16693

(32) A. Montreux, F. Petit and M. Blanchard, Tetrahedron Letters, (1978) 4967

(33) T. J. Katz and J. McGinnis, J. Am. Chem. Soc. **97** (1975) 1592

(34) R. R. Schrock, J. S. Murdzek, J. H. Freudenberger, M. R. Churchill and J.W. Ziller, Organometallics **5** (1986) 25

(35) J. Sancho and R. R. Schrock, J. Mol. Cat. **15** (1982) 75

(36) R. R. Schrock, S. F. Pedersen, M. R. Churchill and J.W. Ziller, Organometallics **3** (1984) 1574

(37) M. R. Churchill, J.W. Ziller, S. F. Pedersen and R. R. Schrock, J. Chem. Soc. Chem. Commun. (1984) 485

(38) L. G. McCullough, J. C. Dewan, R. R. Schrock and J. S. Murdzek, J. Am. Chem. Soc. **107** (1985) 5987

(39) H. Strutz, J. C. Dewan and R. R. Schrock, J. Am. Chem. Soc. **107** (1985) 5999

(40) W. E. Buro and M. H. Chisholm, Advances in Orgenomet. Chem. **27** (1988) 311

(41) M. R. Churchill, J.W. Ziller, J. H. Freudenberger and R. R. Schrock, Organometallics **3** (1984) 1554

(42) T. J. Katz and S. I. Lee, J. Am. Chem. Soc. **107** (1985) 5993

(43) K. Weiss and G. Lößel, to be published

Index